Statistics in
Human Genetics and
Molecular Biology

CHAPMAN & HALL/CRC
Texts in Statistical Science Series

Series Editors
Bradley P. Carlin, *University of Minnesota, USA*
Julian J. Faraway, *University of Bath, UK*
Martin Tanner, *Northwestern University, USA*
Jim Zidek, *University of British Columbia, Canada*

Texts in Statistical Science

Statistics in Human Genetics and Molecular Biology

Cavan Reilly

University of Minnesota
Minneapolis, Minnesota, U.S.A.

CRC Press
Taylor & Francis Group
Boca Raton   London   New York

CRC Press is an imprint of the
Taylor & Francis Group an **informa** business

A CHAPMAN & HALL BOOK

First published 2009 by Chapman & Hall/CRC

Published 2019 by CRC Press
Taylor & Francis Group
6000 Broken Sound Parkway NW, Suite 300
Boca Raton, FL 33487-2742

© 2009 by Taylor and Francis Group, LLC
CRC Press is an imprint of Taylor & Francis Group, an informa business

No claim to original U.S. Government works

ISBN-13: 978-1-4200-7263-1 (hbk)

Library of Congress Cataloging-in-Publication Data

Reilly, Cavan.
    Statistics in human genetics and molecular biology / Cavan Reilly.
       p. cm. -- (CHAPMAN & HALL/CRC texts in statistical science series)
    Includes bibliographical references and index.
    ISBN 978-1-4200-7263-1 (alk. paper)
       1. Human genetics--Statistical methods. 2. Genomics--Statistical methods. 3. Molecular biology--Statistical methods. I. Title. II. Series.

QH438.4.S73R45 2009
572.8072'7--dc22
                                                 2009014171

Visit the Taylor & Francis Web site at
http://www.taylorandfrancis.com

and the CRC Press Web site at
http://www.crcpress.com

Contents

Preface

This book is intended for students in Statistics, Biostatistics, Computer Science and related fields in applied Mathematics who seek to gain a basic knowledge of some of the problems arising in the analysis of genetics and genomics. As such, it is assumed the reader is familiar with calculus and linear algebra and has been exposed to a course that covers the basics of probability and statistics assuming knowledge of calculus and linear algebra. The unifying theme of all the topics revolves around the questions: "What does this segment of DNA do?" and "Which segment of DNA does this?" The selection of topics is largely based on a course I took while a student at Columbia that was offered by Shaw-Hwa Lo and Herman Chernoff (although they obviously can't be blamed for any of the shortcomings one may notice here). The material of that course was the basis for a course I started in conjunction with Wei Pan and Hegang Chen. Wei and I continued to teach this course and refined the subject matter as we progressed. To all of these people, I am deeply indebted for their help. The largest change that Wei, Hegang and I made to the material that Drs. Lo and Chernoff included in their class is the inclusion of a substantial amount of material on microarrays (which were just emerging when I took Lo and Chernoff's course).

This book is not intended to be the authoritative treatment of the topics it covers, as such a book would be enormous and unsuitable as a text. Rather, the goal here is to introduce the student to a diverse set of problems and a number of approaches that have been used to address these problems. In the process, it is hoped that the reader will learn enough Biology, Statistics, Computer Science and Bioinformatics so that when confronted with a new problem at the interface of these areas, the reader will have some ideas regarding what will likely constitute a reasonable approach. This seems preferable to covering this material in an encyclopedic fashion since technology is constantly changing. Indeed, even over the course of time in which this book was drafted, technological developments have altered our view regarding optimal approaches to certain problems. For example, the role of linkage studies relative to association studies for disease gene mapping has changed substantially over the last several years (and as I write this, some are questioning the role of association analysis relative to sequencing entire genomes of carefully selected subjects). Nonetheless, topics were selected that will likely have value as technology changes. For example, classification and cluster analysis will still be important tools when we master quantification of proteins or metabolites. Other topics were selected because they illustrate the use of methods that are widespread among researchers who analyze genomic data, such as hidden Markov models and the extreme value distribution.

In addition to those already mentioned, I would like to thank all the students who attended this course and provided valuable feedback (even if some of this feedback just consisted of an odd facial expression betraying confusion, likely due to my imperfect presentation). In addition John Connett (the Division head while I drafted this book) deserves thanks for understanding that writing a book is a huge undertaking that necessarily detracts from time I could be doing other things. Thanks also to Julie Frommes and my family. Finally thanks to Rob Calver and all the people at Chapman & Hall/CRC Press for their assistance and patience.

Cavan Reilly
Minneapolis, MN
February, 2009

CHAPTER 1

Basic Molecular Biology for Statistical Genetics and Genomics

1.1 Mendelian genetics

Although it has been recognized for thousands of years that traits are passed from organisms to their offspring, the science of genetics began with the experiments of Gregor Mendel. By crossing various strains of peas, Mendel was able to deduce several principles of genetics. The most basic is *unit inheritance*, which states that inheritance is determined by discrete quantities known as *genes* and genes don't mix. For example, when Mendel crossed round peas (i.e. peas with a smooth surface) and wrinkled peas, all of the offspring were round, the offspring were not "kind of wrinkled." In this example, we think there is some gene which controls if the pea is wrinkled or not, and the gene takes 2 values; round and wrinkled. We call the different values taken by the gene *alleles*. Peas (and humans) have 2 copies of all genes, one from each parent. The set of 2 alleles of a gene for an individual is known as the *genotype* of the individual. In contrast, an observable feature of an organism is referred to as the *phenotype* of the individual. For example, if a pea plant has the allele for round (denoted R) and wrinkled (denoted r) at the pea shape gene, the genotype is Rr (or rR, the order of the symbols is irrelevant). If a pea plant has this genotype, it will appear round, hence we say it has the round phenotype. Another way to think about the relationship between phenotype and genotype is that genotypes help determine the phenotype (in conjunction with environmental effects). The term genotype is also used to refer to the alleles an individual has at more than one gene. If we also know which pairs of alleles were inherited from which parent, then we know the individual's *haplotype*. For example, suppose there are 2 genes, each denoted by a distinct letter, and each with 2 alleles, which we will denote Aa and Bb. If someone has the genotype A, A and B, b then we know one parent must have had the pair of alleles A, B and the other parent must have had the alleles A, b, hence in this case we know haplotype of this individual. In contrast, if the genotype had been A, a and B, b then the parents could have 2 possible configurations for their alleles: A, B and a, b or A, b and a, B. In the latter case we cannot deduce the haplotype from the genotype.

If the 2 alleles for a gene are distinct, we say that individual is *heterozygous*, while if the 2 alleles are the same, we say the individual is *homozygous*. For many traits, homozygous genotypes translate into definite phenotypes. For the pea example, if the genotype is rr, the pea will be wrinkled (i.e. the wrinkled phenotype) and if the genotype is RR the phenotype is round. The phenotypes of heterozygous individuals are more complicated. In classical Mendelian genetics, there are 2 ways in which a heterozygous genotype translates into a phenotype and this mapping of genotype to phenotype depends on the trait. For the pea example, the Rr genotype corresponds to a round phenotype, hence we say round is the *dominant* trait and wrinkled is the *recessive* trait. For some traits, heterozygous genotypes imply the individual displays both phenotypes. Such traits are called *codominant*. From experimentation, Mendel also deduced that alleles *segregate randomly*, that is, if an individual has 2 distinct alleles for some gene then an offspring of this individual is equally likely to obtain each allele. Mendel's final deduction was that different genes combine independently of one another (this is the principle of *independent assortment*). We will see that Mendel's principles, while useful, are not entirely correct. In particular, the principle of independent assortment is only correct for certain traits. Traits that obey Mendel's first 2 principles are referred to as *Mendelian traits*, and several hundred medical disorders are well modeled as Mendelian traits (i.e. attributable to a single gene which is either dominant or recessive). Cystic fibrosis is a noteworthy example as are some common forms of hearing loss. Many other traits (e.g. height) are far too complex to fit neatly into the Mendelian framework, hence such traits are referred to as *complex traits*.

1.2 Cell biology

While Mendel's insights regarding the regulation of phenotypes via some unobserved mechanism that is transmitted from parents to offspring was fundamental to our understanding of how some traits are regulated, it left much unexplained. Another fundamental insight into the creation and maintenance life was the observation of cells in living organisms. While cells had been described as early as the mid-1600s, developments in microscopy during the nineteenth century allowed biologists to learn that living organisms are composed of minute, distinct units that have a membrane and structures inside this membrane immersed in a fluid. The structures are called *organelles*, the fluid is called *cytosol* and the collection of organelles and fluid is referred to as the *cytoplasm*. While different types of cells have different organelles, some organelles are common to many cells among complex organisms. For example, mitochondria are organelles found in most human cells that are used by the cell to generate energy so the cell can carry out its functions. Complex organisms, such as humans, are estimated to be composed of trillions of cells. Cells are frequently organized into types, such as the familiar red blood cell, and many types (over 200 currently in humans) have been distinguished. Cells are generally distinguished by the function they carry out in the organism: for example, red blood cells allow for the transportation of oxygen and carbon dioxide throughout the body since these gases will bind to the cell. Complex organisms have special cells, *germ cells*, devoted to the task of reproduction. While there are many single celled organisms, we will focus on multicellular organisms.

While the contents of a cell are inside a membrane, this membrane is semipermeable so as to allow the cell to interact with other cells and its environment. Inside a cell many chemicals are present that interact with each other in a set of extremely complex biochemical reactions. These reactions might cause the cell to move or to secrete some chemical into its cellular environment. In addition, due to external sources, the cell might alter the set of reactions taking place in the cell or in some smaller part of the cell. In this sense, a living organism can be thought of as a collection of intertwined biochemical reactions. Which chemicals are present in a cell and their relative quantity are determined by the action of genes. Almost all human cells have a structure inside the cell called the *nucleus* that holds information on the set of alleles for each gene that an individual has received from its parents. All cells are created by the separation of some parent cell, which is called *cell division*. Cells undergo multiple rounds of cell division, and the process from creation of a new cell to when that new cell divides is called the *cell cycle*. As a cell proceeds through the cell cycle there are alterations that the cell undergoes that are visible with a microscope. Sometimes mistakes occur during cell division which cause the offspring cells to have slightly different alleles from the parent cell. Such mistakes are called *mutations* and are a critical feature of natural selection.

1.3 Genes and chromosomes

Complex organisms have tens of thousands of genes. These genes are mostly located on *chromosomes*, and the spots on the chromosomes where genes are located are referred to as *loci* (locus is the singular). One can observe chromosomes under a microscope in a cell that is dividing. Chromosomes are segments of deoxyribonucleic acid (DNA), and are contained in the nucleus of the cells of eukaryotic organisms. A eukaryotic organism is one whose cells have nuclei, for example, algae, yeast, plants and animals. The other major division of life are referred to as prokaryotes, with the most notable example being bacteria. While most tend to think of bacteria as organisms that are harmful to animals and plants, animal tissue (including humans) is full of bacteria that play an important role in physiology. The nucleus is surrounded by a semipermeable membrane called the nuclear envelope that regulates what compounds enter and exit the nucleus. In addition to the nuclear DNA organized into chromosomes, in humans, some DNA is found in the mitochondria. Most genetic investigations ignore this mitochondrial DNA since there are only 37 genes in mitochondrial DNA whereas there are 30,000-35,000 genes on the chromosomes (the exact number is not yet known). The 37 mitochondrial genes are involved in energy synthesis inside the mitochondria, and other genes located on the chromosomes that reside in the nucleus are involved with this process too.

Humans have 46 chromosomes, arranged in 23 pairs of chromosomes. Of these 23 pair, 1 pair is known as the sex chromosomes and the other 22 are called *autosomal* chromosomes. The centromere is a location near the center of a chromosome that plays an important role during cell division and lacks genes. The centromere can be observed during certain stages of a cell's development, and the shorter end of the chromosome is referred to as the p arm while the longer end is called the q arm. The ends of the arms of the chromosomes are referred to as *telomeres.*

Organisms receive 1 chromosome of each pair from each parent, hence, for autosomal chromosomes, the pairs all have the same gene loci, but possibly different alleles. An individual also receives one of the sex chromosomes from each parent, but these are not necessarily pairs. There are 2 sorts of sex chromosomes, denoted X and Y, so there are a total of 24 distinct human chromosomes. An individual with 2 X chromosomes is a female (only one of these is used by cells, whereas the other is inactivated), while an individual with an X and a Y chromosome is a male. Since females have 2 X chromosomes, they always transmit an X chromosome to their offspring, while males contribute either an X or a Y chromosome (which is actually transmitted depends on chance), hence it is the sperm cell which determines the sex of the offspring. There are several serious medical disorders attributable to a gene which resides on the X chromosome: such disorders are referred to as *X-linked* disorders (or X-linked traits more generally). Such disorders are recessive as only one disease allele results in the disease trait. These traits can seem to skip a generation since an affected mother may carry a disease allele on the disactivated X chromosome and hence be unaffected (color blindness is an example of this phenomenon). A woman with a disease allele on an X chromosome can nonetheless transmit the disease allele to her offspring, and if the offspring is a male he will only have the disease allele at this locus and thereby be affected. Humans receive all of their mitochondrial DNA from the maternal source, hence traits determined by mitochondrial genes can exhibit transmission patterns similar to X-linked traits.

The autosomal chromosomes are numbered according to their length, with chromosome 1 being the longest, and the sex chromosomes are chromosome pair 23. A surprising feature of chromosomes in higher organisms (like humans) is that most of the DNA is not part of a gene (over 95% for humans). In addition, most genes are not composed of contiguous segments of DNA, instead there are interruptions in the coding of a gene. Such interruptions are known as *introns,* while the segments of DNA which encode a gene are called *exons.* Some genes can be used by the cell in more than one way since the exons can be assembled in a variety of ways, thereby generating different gene products from the constituent parts of a single gene. This phenomenon is referred to as *alternative splicing* for reasons that will become clear when we discuss RNA. It is currently not entirely clear what is the purpose of this extragenic DNA, but at least some portions of this DNA regulates the working of the genes. Some of the extragenic DNA was likely deposited by a virus in some ancestor of contemporary humans, and these sequences have been passed down through time since they didn't have a harmful effect on this ancestor. Often the term locus is used to refer to a location on a chromosome even if there is no gene there, and similarly, allele is used to refer to the segment of DNA that an individual has at the locus. Since chromosomes are composed of DNA, further discussion of DNA is necessary.

1.4 DNA

The status of human cells is in large measure regulated by DNA. Cells use DNA to make proteins, and the resulting proteins carry out the functions necessary for life. Proteins are built up from a collection of amino acids (there are 20 amino acids), but often incorporate other molecules in their final form (for example, metals and lipids). DNA is composed of a sequence of sugar molecules (2-deoxyribose) joined by phosphate groups. These sugar molecules contain 5 carbon atoms arranged in a ring, and these are referred to by a number between 1' and 5'. To each sugar molecule, at the carbon atom numbered 1', is attached one of four bases; adenine (A), guanine (G), cytosine (C) and thymine (T). The bases A and G are referred to as *purines* while C and T are called pyrimidines (due to details regarding the chemical structure of the base). All genetic information is encoded in these 4 bases. A *nucleotide* is a combination of a sugar, a phosphate and a base. Phosphate groups attach to the carbon atoms numbered 3' and 5'. The linear chain of DNA is built by phosphate groups that link the 5' carbon atom of one sugar to the 3' carbon atom of the next sugar. The bases from distinct nucleotides tend to form hydrogen bonds with one another in a specific manner; A binds with T and G binds with C. Due to this pairing, for every sequence of DNA there is a complementary sequence which is defined by substituting a T for an A, an A for a T, a G for a C and a C for a G.

DNA has a tendency to bind to its complementary copy and form a double helix (i.e. a twisted ladder configuration), but the hydrogen bonds which form between the nucleotides from different strands (i.e. the rungs of the twisted ladder) are easily broken (which is necessary for DNA to fulfill its function). While the cell is not dividing (a period of the cell cycle known as interphase), the DNA molecule actually exists as a complex (i.e. a large molecule made up of a number of molecules) with a group of proteins called histones and a group of acidic nonhistone proteins; this complex is referred to as *chromatin*. Basically, the DNA winds around the histone complex, and is further condensed into a structure called a solenoid; these structures are far more compact than free DNA and help protect it from the sometimes harsh environs of the cell. In a DNA molecule, the nucleotides are arranged in a linear fashion along a chemical backbone composed of the alternating phosphates and sugars. If you start reading the bases as you travel along a DNA molecule, then groups of 3 nucleotides are referred to as *codons*. Of course which end you start reading from makes a difference. Recall the carbon atoms are numbered with connections from carbon 5' to carbon 3', so the ends of a DNA segment are referred to as 3' and 5', and you read codons by starting at the 5' end. Moreover, in the double helical formation in which DNA resides, the 2 strands have a reverse orientation in terms of this 5' to 3' directionality. Note that genes are located on both strands and can even overlap in the sense that there is a gene on both of the strands in the same location (although this is thought to be rare). We use the term *genome* to refer to the collection of DNA of some organism. The human genome consists of approximately 3 billion basepairs, and about 99.9% of these basepairs are common among humans. Thus, only about 3 million basepairs are responsible for genetic differences among people. Environmental sources (e.g. diet and climate) are also important factors that gives rise to differences observed among humans. Moreover, there can be interactions between genetic factors and environmental factors that impact the phenotype of an organism.

1.5 RNA

The process whereby DNA is used to make proteins is composed of several steps. Proteins are produced by ribosomes, which are organelles found outside the nucleus of the (eukaryotic) cell (prokaryotes also use ribosomes in this fashion), hence there needs to be a way of moving the information contained in DNA from the nucleus to the ribosome. Ribonucleic acid (RNA) is the molecule which allows for this communication. RNA is much like DNA, except the sugar group is ribose instead of 2-deoxyribose and uracil (U) takes the place of thymine (T). These chemical differences between DNA and RNA imply that RNA doesn't have a strong tendency to form a double helix, unlike DNA. The first step in generating a protein from DNA is when premessenger RNA makes a copy of a DNA segment after the hydrogen bonds that hold the double helix of the double stranded DNA molecule are separated near a gene (the breaking of hydrogen bonds is regulated by proteins called *transcription factors*). After this copying (known as *transcription*), some editing of the segment takes place. Certain sequences of nucleotides indicate that a gene is about to start or end (called start and stop codons), so the editing deletes the portions of DNA which are not part of a gene (this is why the alternative splicing terminology is used for some genes). In addition to this editing, a cap structure is attached on the 5' end of the sequence and a poly-A tail (i.e. a variable length sequence of adenine nucleotides) is added to the 3' end. These structures help stabilize the resulting molecule since they can prevent the degradation of the molecule by allowing certain proteins to bind to its 2 ends. This editing and attachment of structures to the ends of the RNA molecule yields messenger RNA (mRNA) which can then pass through the nuclear envelope and potentially associate with ribosomes outside of the nucleus. When an mRNA molecule interacts with a ribosome, transfer RNA (tRNA) helps coordinate amino acids so that a chain based on the sequence of nucleotides present in the mRNA is built (this process is known as *translation*). For each codon in the mRNA molecule, a particular amino acid is attached to the growing chain of amino acids at the ribosome (the collection of naturally occurring amino acids is presented in Table 1.1). In this way a chain of amino acids is built up by the ribosome forming a *peptide* (a sequence of amino acids). The process whereby DNA is used to produce a protein or active mRNA molecule is referred to as *gene expression*. The mRNA molecule is released when translation is complete, at which point it may associate with proteins that degrade the mRNA molecule, or it may interact with another ribosome and produce more protein. When there is high demand for a gene product, several ribosomes may associate with each other to produce protein from the same mRNA molecule, thereby generating what is called a polyribosome. While we can think of DNA as a linear polymer, RNA has a tendency to fold into more complicated structures, and these structures impact the functioning of the RNA molecule.

Table 1.1 *The naturally occurring amino acids, with abbreviations and single letter designators.*

Name	abbreviation	designator
Aspartic Acid	Asp	D
Glutamic Acid	Glu	E
Lysine	Lys	K
Arginine	Arg	R
Histidine	His	H
Asparagine	Asn	N
Glutamine	Gln	Q
Serine	Ser	S
Threonine	Thr	T
Tyrosine	Tyr	Y
Cysteine	Cys	C
Glycine	Gly	G
Alanine	Ala	A
Valine	Val	V
Leucine	Leu	L
Isoleucine	Ile	I
Proline	Pro	P
Methionine	Met	M
Phenylalanine	Phe	F
Tryptophan	Trp	W

1.6 Proteins

Proteins are molecules composed mostly of amino acids and are created from RNA via a translation system that uses the *genetic code*. While the genetic code is not entirely universal, for example, it is slightly different for human mitochondrial DNA, it is nearly universal for nuclear DNA. The genetic code describes how codons get mapped to amino acids. Since there are 4 distinct nucleotides, there are $4^3 = 64$ possible codons. Given a DNA molecule, if one knows which base is the first base in a codon, then one know the reading frame for that molecule and can deduce the sequence of codons. As noted above, in the process of creating a protein, each codon gets transcribed into one amino acid. Each of the 20 amino acids has a name (e.g. tyrosine), an abbreviated name (e.g. for tyrosine, it is Tyr), and a single letter name (e.g. for tyrosine this is Y). Since there are 64 possible codons and only 20 amino acids, there is some redundancy in the genetic code. This redundancy is not completely without a pattern; as seen in Table 1.2 it is least sensitive to the nucleotide in the third base and most sensitive to the nucleotide in the first. The start codon is AUG, so all proteins start with the amino acid methionine (whose symbol is M), but this M is sometimes cleaved once the protein has been assembled. When a stop

Table 1.2 *The genetic code. The amino acid associated with each codon is indicated by ts single letter name and is presented to the right of the codon. An X indicates a stop codon and the start codon is AUG.*

codon	amino acid	codon	amino acid	codon	amino acid	codon	amino acid
UUU	F	CUU	L	AUU	I	GUU	V
UUC	F	CUC	L	AUC	I	GUC	V
UUA	L	CUA	L	AUA	I	GUA	V
UUG	L	CUG	L	AUG	M	GUG	V
UCU	S	CCU	P	ACU	T	GCU	A
UCC	S	CCC	P	ACC	T	GCC	A
UCA	S	CCA	P	ACA	T	GCA	A
UCG	S	CCG	P	ACG	T	GCG	A
UAU	Y	CAU	H	AAU	N	GAU	D
UAC	Y	CAC	H	AAC	N	GAC	D
UAA	X	CAA	Q	AAA	K	GAA	E
UAG	X	CAG	Q	AAG	K	GAG	E
UGU	C	CGU	R	AGU	S	GGU	G
UGC	C	CGC	R	AGC	S	GGC	G
UGA	X	CGA	R	AGA	R	GGA	G
UGG	W	CGG	R	AGG	R	GUG	G

codon is reached (represented by X in Table 1.2), translation stops, and the protein is released into the cytoplasm of the cell. At that point, proteins undergo changes in shape (conformational changes) and may incorporate other molecules so that the protein can serve its function. Some proteins change their shape depending on the environment, but all seem to undergo immediate changes once released from the ribosome. In addition, proteins can undergo *post-translational modifications*. These are chemical alterations wrought by other proteins on an existing protein. Unlike DNA (which we think of as a linear segment of bases), the actual shape of a protein has a tremendous impact on the function of the protein, and the final shape in a particular environment is not easily deduced from the sequence of amino acids, although certain rules do seem to hold. The term polymer is used to describe molecules that are composed of a sequence of identical subunits, and the term *biopolymer* is used to refer to a polymer that exists in living things, such as DNA (whose subunits are nucleotides), and proteins (whose subunits are amino acids).

An amino acid is composed of an amino group (NH_2), a carboxyl group (COOH), a hydrogen atom, and a residue, and these bind at a common carbon atom (denoted C_α). The particular residue determines the particular amino acid. There are 4 groups of amino acids that are commonly distinguished: basic (K, R and H), acidic (D and E), uncharged polar (N, Q, S, T, Y and C) and nonpolar neutral (the remainder). Amino acids within a group tend to behave in the same fashion, for example, nonpolar neutral amino acids are hydrophobic, meaning they tend to avoid association with water molecules. Each amino acid has a unique set of chemical properties. For example, strong chemical bonds (called disulphide bonds, or disulphide bridges) tend to form between cysteine residues common to a single protein (or across proteins), and these bonds can impact the final shaped of the protein. A peptide chain grows by forming bonds (known as *peptide bonds*) between the amino group of one amino acid and the carboxyl group of the adjacent amino acid (giving off water). These peptide bonds form a backbone along which the residues reside. All of the atoms in the peptide bond (NH-CO) lie in the same plane, so the 3 dimensional structure of a peptide depends on only 2 angles per C_α. There are only certain combinations of these 2 angles which are allowed due to the residues interfering with one another in space (which are referred to as *steric constraints*), and these constraints give rise to the commonly encountered 3 dimensional structures in proteins. Glycine is largely free of steric constraints because its residue is so small (it is a single hydrogen atom). Two major structures built from the backbone of peptide chains (referred to as *secondary structures*) are commonly encountered; α-helixes (which are helixes of residues) and β-sheets (in which residues line up roughly co-linearly to produce a sheet of residues). There are also structures composed from these secondary structures which have been documented (referred to as *tertiary structures*). The 3 dimensional structure of a protein is important for its functioning since proteins are the working parts of cells, hence the structure of a protein is generally conserved in evolution. Since glycine is not as constrained in its angles, it is often in tight turns in the 3 dimensional structure, hence its location needs to be conserved in random mutations if the protein is to continue its function. Other residues are important for forming certain other 3-dimensional structures, hence they need to be preferentially conserved in mutation too.

1.6.1 Protein pathways and interactions

Many proteins have an effect on the cell by binding to other molecules, such as DNA or other proteins, and the ability of a protein to bind to another molecule is greatly influenced by the 3 dimensional shape of the protein. The object to which a protein binds is called its *ligand*. After a protein binds to another molecule, the other molecule (and the protein itself) may undergo changes. We often refer to this process as one protein *activating* the other since frequently proteins exist freely in the cell until encountering an enzyme which promotes a chemical reaction by lowering the level of energy needed for the reaction to occur.

Proteins are the primary chemicals that are involved in the biochemical reactions described in Section 1.2. For a multicellular organism to survive, the component cells need to coordinate their activities. Such coordination is achieved by intercellular signaling. A cell will respond to certain stimuli (such as heat or lack of nutrients) by producing signaling molecules and releasing these molecules through its cell membrane. This is achieved by a sequence of chemical reactions that involves converting certain segments of DNA into RNA which then allows certain protein products to be produced. These proteins then allow the signaling molecules to bud off the surface of the cell. Animal cells are covered with molecules that selectively bind to specific signaling molecules. If the ligand of a signaling molecule is encountered on the surface of another cell, the signaling molecule will bind to that cell and the cell to which it binds will then undergo a set of changes. Other signaling molecules are able to penetrate the cell membrane. These changes are brought about by the physical binding of the signaling molecule to the surface of the cell as such physical binding can lead the cell to alter the set of proteins that the cell is producing. In this way, changes in gene expression in one cell can lead to changes in the gene expression of many cells. Moreover, as gene expression changes in the second set of cells, they alter the set of proteins they secrete which can lead to other outcomes further down stream in a connected sequence of events. In a given cell, many pathways will operate simultaneously (some involving the same genes) and the outcome of all these forces determines what the cell does. In this way, a cell acts as a certain cell type in response to the genes that it is expressing. Hence it is the action and use of genes via the process of gene expression that determines the phenotype of a cell, and the collective action of cells, which is also coordinated by gene expression, determines the phenotype of an organism.

Proteins are frequently organized into families of proteins based on function. While a variety of methods to construct such families have been proposed, there are typically thousands of families and frequently there are subfamilies distinguished with these families. Databases have been constructed to allow researchers to ascertain the family of a given protein (for example the PANTHER data base of Mi et al., 2006). For example, Eukaryotic protein kinases is one of the largest families of proteins. All proteins in this family have nearly the same sequence, structure and function (they are used by cells to communicate with each other). Protein families typically extend across organisms and frequently organisms will have multiple versions of proteins from a family. Another commonly studied gene family is the major histocompatability complex. Three distinct classes of this protein family are commonly distinguished, and they are called class 1, 2, and 3. This family of proteins is central to immune control. Humans have 3 proteins that belong to MHC class 1, and they are called the human leukocyte antigens, which, like most proteins and genes, is usually referred to by its abbreviated name HLA (the 3 different types in humans are called HLA-A, HLA-B and HLA-C). These genes produce receptors that cover cell surfaces and allow cells from the immune system to determine if other molecules are foreign or produced by itself.

A major initiative aimed at classifying proteins (and thereby the genes that produce these proteins) is the *Gene Ontology Consortium*. This is a group of researchers attempting to develop a common set of procedures for classifying proteins, and the classification system is referred to as the gene ontology (or just GO). A distinctive feature of GO is that it classifies genes by 3 separate criteria: biological process, cellular component and molecular function. These ontologies attempt to place genes into a directed acyclical graph so that as one moves further down the graph structure there are fewer genes in any category and the process, component or function is more precise, i.e. the classifications get more coarse as one moves up the graphical structure. A gene in one group at a fine level of separation can point to multiple coarser classifications since a gene may play a different role in a different cellular context. Proteins that are involved in the same cellular process are typically referred to as being in a common pathway. Note that there is not a simple relationship between gene families and pathway membership as genes from the same family may or may not be in a common pathway.

As an example of a well understood pathway, in the growth factor-regulated MAP kinase pathway (Robinson and Cobb, 1977), once the signaling molecule (a growth factor) binds to the cell surface receptor for a tyrosine kinase (an enzyme used for intracellular communication and a member of the Eukaryotic protein kinase family), the receptor and other proteins nearby undergo post-translational modifications that lead to certain other proteins coming to the location where the receptor is located. These other proteins include Ras (an important protein in the development of cancer) which then becomes activated. Once activated, Ras can recruit the protein Raf to its location which then activates the protein MEK, which in turn activates MAP kinase. After MAP kinase is activated, it can activate transcription factors which then alter the pattern of gene expression (activated MEK also activates these other transcription factors). For this reason, researchers often think of genes as working in groups, or as part of a biological pathway. In actuality, many receptors have their ligands bind to a cell, and the response of the cell to the collection of these signals depends on the current levels of signals and the levels of all proteins in the cell. For this reason, it makes sense to frame many questions about the operation of proteins in cells in terms of the interactions between pathways of genes.

1.7 Some basic laboratory techniques

There are a number of fairly standard techniques that can be used to measure the level of a DNA, RNA or protein molecule from a collection of cells. A basic distinction in these methods is between cloning and hybridization. In DNA cloning, a particular DNA fragment is amplified from a sample to produce many copies of the DNA fragment. After this cloning, the amplified products are separated from the rest and properties of the fragment can be studied. The most commonly used method for DNA cloning is called polymerase chain reaction (PCR). This technique can also be used to measure the quantity of RNA by first using the RNA to make cDNA (i.e. a sequence of DNA that is the complement of the DNA sequence used to generate the RNA molecule, called complementary DNA), then using PCR to amplify this cDNA (this is called reverse transcriptase PCR because the enzyme reverse transcriptase is used to generate the cDNA from the RNA).

In contrast to cloning based techniques, with molecular hybridization, a particular

DNA or RNA fragment is detected by another molecule. An example of the latter technique, referred to as a southern blot hybridization, can be used to determine aspects of a sequence of DNA at specific locations. In particular this method can be used to determine what alleles an individual has at a particular locus. Northern blots are a variation of southern blots that allow one to measure the quantity of RNA in a collection of cells. Finally, a slightly different technique, called a western blot, can be used to quantify the level of a protein in a particular tissue sample. All of these techniques require one to specify the exact molecule (i.e. DNA fragment or protein) one wants to measure and typically only a few types of molecules will be quantified in a study that uses these techniques.

In the final chapters we will discuss microarray technology, a molecular hybridization technique, which allows one to measure the level of mRNA for thousands of genes simultaneously. Consideration of thousands of genes instead of just a handful has changed the nature of investigations in molecular biology to the extent that researchers refer to considerations of just a few genes as genetics, whereas consideration of thousands of genes is referred to as *genomics*. Nonetheless, these techniques can be used to investigate if a gene or protein is expressed at different levels in different patient populations or in different tissue types. Since we generally think of DNA as being the same in every cell in the body (which isn't quite true since there are mutations that occur during cell division, and some of these are extremely important since such mutations may induce cancer), usually DNA is isolated from white blood cells (i.e. cells that function as part of the immune system and are easily found in a blood sample) and in most studies involving DNA, samples of white blood cells are stored in special freezers so one can further investigate the DNA in the future. While red blood cells may appear to be a more natural choice (since they are the most common cell in the blood) they don't have a nucleus since they are terminally differentiated cells (i.e. mature red blood cells don't divide).

There are also a number of organisms that are frequently studied in genetics. The use of another animal to understand human genetic processes constitutes a model for that genetic process. Studying these organisms can lead to insights not only regarding genetic mechanisms, but also provides a way to develop technology in a more cost effective way than using human samples. The prokaryote *Escherichia coli* (abbreviated *E. coli*) is commonly used to understand some basic genetic mechanisms. Even though humans are distantly related to this bacterium, a number of genetic mechanisms are common. A simple eukaryote that has received a great deal of attention is *Saccharomyces cerevisiae*, a single celled yeast. Surprisingly, a large fraction of yeast genes have a similar gene found in mammals. Other extensively studied organisms include *Caenorhabditis elegens* (a 1 mm long roundworm), and *Brachydanio rerio* (the zebrafish) for insights regarding development and *Drosophilia melanogaster* (the fruit fly). All these organisms have quick developmental cycles and are easy to genetically manipulate. Finally, *Mus musculus* (i.e. the mouse) has been used in many investigations since mice have short life spans and almost every gene has a corresponding human gene (in fact large chromosomal regions are common between humans and mice). The entire genome (i.e. sequence of nucleotides in their DNA) is available for all of these important models for human genetics.

1.8 Bibliographic notes and further reading

While the intention of this chapter is to provide sufficient background material for what is discussed here, there is an enormous amount of further reading one could do to be better equipped to study statistical problems in genetics and molecular biology. There are a good number of excellent texts on genetics, such as Thompson, McInnes and Willard (1991) and Vogel and Motulsky (1997). There are also many good books that cover genetics and cell biology from a more molecular biology perspective, such as Lodish et al. (2001) and Strachan and Read (2004). Most of the material presented in this chapter is from these sources.

1.9 Exercises

1. If 98% of human DNA is not used for producing proteins (i.e. non-coding DNA), what is the mean length of a gene in terms of bases if we ignore the possibility of genes on separate strands overlapping? What is the mean length of a protein in terms of amino acids?

2. If 98% of human DNA is not used for producing proteins (i.e. non-coding DNA), what is the mean number of exons per gene if we ignore the possibility of genes on separate strands overlapping and the mean exon length is 120 bp?

3. In humans, what is the mean number of gene alleles on a chromosome pair if all chromosomes have the same number of genes?

4. If a gene has 4 exons each with 200 codons, what is the length of its associated mRNA molecule that is copied from the DNA molecule? How many introns does it have in its DNA sequence?

5. Suppose nucleotides arise in a DNA sequence independently of each other and all are equally likely. How often does one expect to encounter a start codon just by chance? How often will one encounter a stop codon by chance? What is the mean length of the genes one would detect in such random DNA? How many would you detect in a sequence the length of the human genome?

6. Translate the following mRNA sequence into a sequence of amino acids:
   AUGUUGUGCGGUAGCUGUUAU.

7. Show that 2 unique proteins are encoded by the following DNA sequence:
   UUAGAUGUGUGGUCGAGUCAUACUGACUUGA.
   Note that the gene doesn't necessarily start at the first letter in this sequence, the orientation of the sequence is unknown (i.e. we don't know which end is the 3' or 5' end) and a gene may be found on the complementary sequence. Also, give the protein sequences.

CHAPTER 2

Basics of Likelihood Based Statistics

2.1 Conditional probability and Bayes theorem

Traditionally, statisticians have modeled phenomena by supposing there are population parameters which are fixed at certain unknown values, and then tried to estimate (i.e. give a best guess of) these parameters. In this framework, known as the frequentist framework, you observe data given that the parameter value takes some unknown fixed value. In contrast, Bayesian statisticians treat all quantities, data and parameters, as random variables. In the Bayesian framework, we envision the following process as responsible for the data we observe: the parameter values are drawn from a probability distribution, then we observe the data conditional on the values of the parameters. The problem of statistical inference is how to convert the observations we have conditional on the parameter values into information about the parameters.

Bayes theorem provides a mathematically rigorous way to convert from statements about the distribution of the data given the parameter values into statements about the parameter values conditional on the data. Let θ represent a parameter (or set of parameters) in a statistical model (for example the mean of a distribution) and let y represent the data we have observed (in general y may be a vector, i.e. a collection of measurements). We denote the density of θ by $p(\theta)$ and the conditional distribution of θ given y by $p(\theta|y)$. This notation deviates from the common practice of denoting the density of a random variable by expressions like $p_\theta(y)$ where y is the argument of the density. For example, the density of a normal distribution with mean θ and variance 1 in the latter notation is $p_\theta(y) = \frac{1}{\sqrt{2\pi}} e^{-\frac{1}{2}(y-\theta)^2}$ whereas in the former notation we have $p(y|\theta) = \frac{1}{\sqrt{2\pi}} e^{-\frac{1}{2}(y-\theta)^2}$.

Bayes theorem is derived by 2 applications of the definition of conditional probability:

$$
\begin{aligned}
p(\theta|y) &= \frac{p(\theta, y)}{p(y)} \\
&= \frac{p(y|\theta)p(\theta)}{p(y)}.
\end{aligned}
$$

We call $p(\theta|y)$ the *posterior distribution* of θ given the data y, $p(y|\theta)$ is called the *likelihood* and $p(\theta)$ is called the *prior distribution*. Since $p(\theta|y)$ is a probability density in the argument θ, we know $\int p(\theta|y)\,d\theta = 1$, hence if we have the likelihood and the prior we can, in principal, always compute $p(y)$ by integrating the product of the likelihood and the prior, setting the result equal to one and using this to obtain $p(y)$, but in fact we often don't need $p(y)$ at all.

The prior distribution is chosen to reflect any knowledge one may have about the parameters prior to the current data analysis. Often a great deal of research has already been conducted on a topic, so the prior lets the researcher incorporate this knowledge and sharpen the results already obtained, in contrast to simply replicating results already available in the scientific literature. In other contexts, although there is no previous research on a topic, one can supply a prior distribution. As an example, suppose you are trying to locate the chromosome on which a gene which is responsible for some Mendelian trait resides. In this context, if you have no idea where the gene is, your prior distribution could be summarized by saying you think the gene is equally likely to be on any of the chromosomes. More elaborate priors could be constructed by taking into account that not all chromosomes are the same length or have the same number of genes. The first prior is an example of a *non-informative prior* because the prior distribution merely expresses our total ignorance about the chromosome which has the gene. For continuous parameters, like the mean of a normal distribution, we may suppose that prior to seeing the data, the mean is equally likely to be any real number. We can approximate such a prior with the density $p(\theta) = 1/(2k)$ for $-k \le \theta \le k$ then let k become arbitrarily large.

2.2 Likelihood based inference

When we use a non-informative prior, our posterior is often proportional to the likelihood, hence we can simply use the likelihood to conduct inference when we do not have useful prior information or are reluctant to attempt to incorporate the prior information in our analysis. One constructs the likelihood for a model with reference to a probability model. The likelihood is the joint distribution of the data given the parameter values, so a probability model supplies this joint distribution. Often we have data which we are willing to model as independent measurements drawn from a common distribution, and we refer to such data as being independent and identically distributed (abbreviated as iid). Under the iid assumption we can find the joint distribution of the data given the parameters by multiplying all of the marginal distributions

$$p(y_1, \ldots, y_n|\theta) = \prod_{i=1}^{n} p(y_i|\theta).$$

We sometimes denote the likelihood with $L(y_1, \ldots, y_n|\theta)$ and use $\ell(y_1, \ldots, y_n|\theta)$ to denote its natural logarithm. Often Bayesian methods are questioned because of the need to specify the prior distribution, but the real challenge to successful statistical modeling is the intelligent specification of the likelihood. In particular, the assumption of independence is often the most questionable and important assumption made.

A very commonly used model for representing the number of successes in n independent trials is the binomial model. The binomial model supposes that each of the n trials is either a success or a failure, trials are independent and all have the same probability of success, denoted θ. If $n = 1$ we speak of the Bernoulli distribution with parameter θ. If $y = 0$ when the trial is a failure and $y = 1$ when the trial is a success, we can write the likelihood for y as

$$L(y|\theta) = \theta^y(1 - \theta)^{1-y}$$

since $P(y = 1|\theta) = \theta$ and $P(y = 0|\theta) = 1 - \theta$. For $n > 1$ we need to account for the fact that there are multiple ways to obtain y successes in n failures. In this case we find the expression $L(y|\theta) = \binom{n}{y}\theta^y(1 - \theta)^{n-y}$. While a number of priors are frequently used for this model (and the Beta prior is discussed below) a simple prior here is to assume that θ is uniform on the interval $(0,1)$.

While we can always write down a mathematical expression for the posterior distribution given the 2 key ingredients, how to use this expression to draw useful conclusions is not necessarily so straightforward. One idea is to report the mean of the posterior distribution, or find an interval so that there is a specified probability of the parameter falling in this interval (such intervals are called probability intervals or credible sets). These 2 summaries require integration of the posterior, and generally this integration cannot be done analytically. An alternative is to use the posterior *mode* (i.e. the value of θ so that $p(\theta|y)$ is as large as possible) to summarize the posterior distribution. If we think in terms of our data generation model, if we use θ_0 to denote the value of the parameter which is drawn in the initial draw from the parameter distribution, then there is an important result which states that, under a variety of assumptions, the probability that the posterior mode differs from θ_0 by any (small) amount will become zero as the sample size increases. This result assumes that the observations are independent (although this can be relaxed somewhat), and that the model is *identifiable*. We say that a model (given the data that we are modeling) is identifiable if distinct parameter values give rise to different values for the posterior. Note that without this last assumption the mode is not a well defined object.

2.2.1 The Poisson process as a model for chromosomal breaks

Consider a segment of DNA of length t as a collection of a large number, say N, of individual intervals over which a break can take place (there is not a break at $t = 0$). Furthermore, suppose the probability of a break in each interval is the same and is very small, say $p(N,t)$, and breaks occur in the intervals independently of one another. Here, we need p to be a function of N and t because if there are more locations (i.e. N is increased for fixed t or vice versa) we need to make p smaller if we want to have the same model (since there are more opportunities for breaks). Since the number of breaks is a cumulative sum of iid Bernoulli variables, the distribution of the number of breaks up until location t from some arbitrary zero, y_t, is binomially distributed with parameters $(N, p(N, t))$, that is

$$P(y_t = k) = \binom{N}{k} p(N, t)^k (1 - p(N, t))^{N-k}.$$

If we suppose $p(N, t) = \lambda t / N$, for some λ, and we let N become arbitrarily large, then we find

$$P(y_t = k) = (\lambda t)^k e^{-\lambda t} / k!,$$

that is, the number of breaks up to and including location t is distributed according to the *Poisson distribution* with parameter λt. Since t can vary, but for any t y_t is a Poisson random variable, y_t is known as the *Poisson process*.

2.2.2 Markov chains

In the previous example, y_t represents a collection of random variables depending on t. We say that y_t is parameterized by t and a parameterized collection of random variables is known as a *stochastic process*. We will only consider stochastic processes where the parameter takes values in a countable space (e.g. the positive real numbers). Often we call such processes *discrete time* processes since the parameter is time in many applications. For us, the parameter will be location on the genome relative to some arbitrary zero, but we will still sometimes use the time terminology. A stochastic process has the *Markov property* if the future is conditionally independent of the past given the present. A stochastic process that has the Markov property is called a *Markov process*. Clearly the Poisson process possesses the Markov property since we can construct it as the sum of the number of independent events up to a certain location t. A *Markov chain* is a Markov process for which the random variables take only countably many values. The sample space of the individual random variables of a stochastic process is referred to as the *state space*, so a Markov chain has a countable state space. A Markov chain can have a parameter space that is not countable, but we will only need to consider discrete time Markov chains here. Moreover, we will only consider Markov chains that have *finite* state spaces. Markov chains are useful tools for likelihood based inference because the likelihood for a Markov chain is easy to use and it represents an important departure from the iid assumption that is frequently made in statistics.

To specify a finite state space Markov chain, one needs to specify a distribution for where the chain starts, and a set of conditional probabilities that specify how the chain moves from one state to another. Let y_t represent the value of the Markov chain at time t. The set of conditional probabilities, $P(y_t = i_t | y_{t-1} = i_{t-1})$, where i_j denotes the state at time j, are conventionally stored in a matrix, called the *transition matrix*. The i, j element of the transition matrix gives the probability that the chain moves to state j at the next time given that the chain is in state i at the current time. Note that the rows of such a matrix sum to one. If λ is a vector that holds the probability that the chain starts in each of the states, i.e. the ith element of λ is the probability that the chain starts in state i, and if P is the transition matrix, then $P\lambda$ gives a vector whose i^{th} element is the probability that the chain is in state i at time 1.

Similar reasoning leads one to conclude that $P^n \lambda$ gives a vector whose i^{th} element is the probability that the chain is in state i at time n (P^n is P matrix multiplied with itself n times).

A natural question to ask about Markov chains is what happens if they run for very long periods of time. First suppose that the Markov chain is such that there is not a way to partition the state space into disjoint sets that cannot be reached from one another. Such a Markov chain is said to be *irreducible*. Next suppose that there is not a fixed sequence of states that the chain must pass through. Such a process is said to be *aperiodic*. If there exists a vector, π, such that $\pi P = \pi$, then we call π an *invariant measure*. If the elements of π sum to one then we call π an *invariant distribution*. For finite state space Markov chains, there always exists an invariant distribution: it is the left eigenvector associated with the eigenvalue 1 of the transition matrix. (A basic result from linear algebra is that transition matrices have an eigenvalue of 1 and all other eigenvalues are less than one.) The important theorem on convergence can now be stated: if the chain is irreducible, aperiodic and has an invariant distribution π, then $P(y_t = j) \to \pi_j$ for any initial distribution λ as $t \to \infty$. For Markov chains with countable state space one must show that there exists an invariant measure in order to demonstrate convergence. There is also a sort of law of large numbers for Markov chains, but such theorems are referred to as ergodic theorems in the Markov process context. For a very nice treatment of these topics (with accessible proofs) consult Norris (1997).

2.2.3 Poisson process continued

Now suppose we have counts y_i of the number of breaks occurring in a chromosome due to some treatment for n distinct individuals. If we want to say something about the typical number of breaks occurring, we could specify a probability model for the process and conduct inference for that model. Given our previously developed model for breaks along a chromosome, we will suppose that y_i for $i = 1, \ldots, n$ given θ are distributed according to a Poisson distribution with parameter θ and the y_i are all independent. We are supposing that all subjects have equal length chromosomes hence all subjects have the same rate parameter θ. While we'll see in Chapter 3 that this is not strictly true, it is approximately correct. This implies that

$$p(y_i|\theta) = \frac{\theta^{y_i}}{y_i!} e^{-\theta}.$$

Since the y_i are mutually independent. the likelihood is given by the expression

$$
\begin{aligned}
L(y_1, \ldots, y_n|\theta) &= \prod_{i=1}^{n} \frac{\theta^{y_i}}{y_i!} e^{-\theta} \\
&= \frac{\theta^{\sum_{i=1}^{n} y_i}}{\prod_{i=1}^{n} y_i!} e^{-n\theta}
\end{aligned}
$$

so that the posterior distribution is

$$p(\theta|y_1,\ldots,y_n) \;=\; \frac{\theta^{\sum_{i=1}^{n} y_i} e^{-n\theta} p(\theta)}{\prod_{i=1}^{n} y_i! p(y_1,\ldots,y_n)}, \tag{2.1}$$

where $p(y_1,\ldots,y_n)$ is the marginal distribution of the data and $p(\theta)$ is the prior on θ. If we now suppose that $p(\theta) \propto 1$ for all $\theta \geq 0$, then we can write our expression for the posterior distribution as

$$p(\theta|y_1,\ldots,y_n) = C(y_1,\ldots,y_n)\theta^{\sum_{i=1}^{n} y_i} e^{-n\theta}$$

where $C(y_1,\ldots,y_n)$ is just a function of the data. We can find an expression for $C(y_1,\ldots,y_n)$ by using the fact that $\int_0^\infty p(\theta|y_1,\ldots,y_n)\,d\theta = 1$, since if we substitute the previous expression for the posterior in this equality we find

$$\begin{aligned}
1 &= \int_0^\infty p(\theta|y_1,\ldots,y_n)\,d\theta \\
&= C(y_1,\ldots,y_n)\int_0^\infty \theta^{\sum_{i=1}^{n} y_i} e^{-n\theta}\,d\theta \\
&= C(y_1,\ldots,y_n)\frac{\Gamma(\sum_{i=1}^{n} y_i + 1)}{n^{\sum_{i=1}^{n} y_i+1}}
\end{aligned}$$

where we have used the definition of the gamma function, $\Gamma(x) = \int_0^\infty u^{x-1}e^{-u}\,du$ (recall that $\Gamma(x) = (x-1)!$, a result that follows from integration by parts in the integral definition of the gamma function). Note that although we are able to obtain an expression for the marginal distribution of the data in this example, in general we may not be able to obtain an explicit expression for this quantity. Thus we find that the posterior distribution is given by

$$p(\theta|y_1,\ldots,y_n) = \frac{n^{\sum_{i=1}^{n} y_i+1}}{(\sum_{i=1}^{n} y_i)!}\theta^{\sum_{i=1}^{n} y_i} e^{-n\theta}.$$

If we think of this expression as a probability distribution with argument θ, we recognize this distribution as the gamma distribution with parameters $\sum_i y_i + 1$ and n. Hence we can summarize the posterior distribution by finding the mean and variance of the gamma distribution and summarize the posterior via these 2 moments. In this case, the posterior mean is $\int_0^\infty \theta p(\theta|y_1,\ldots,y_n)\,d\theta = \bar{y} + \frac{1}{n}$ and similarly the posterior variance is $\frac{\bar{y}}{n} + \frac{1}{n^2}$.

When the sample size is not large and the posterior is either not unimodal or heavily skewed, the posterior mode may not be a useful summary of the posterior distribution. As an alternative, we could use a computer to simulate many draws from the gamma distribution with these parameters (most software packages can draw samples from this distribution), then plot a histogram of the simulated draws. This simulation based approach to summarizing a posterior distribution is the most common method for conducting inference in a Bayesian setting. While simulating from the posterior distribution is easy in this case (because the posterior has a standard form), frequently this can be difficult. Fortunately a number of general methods for drawing samples from a probability distribution have been developed. We will discuss some of these methods in Section 2.6. Simulation provides a very powerful approach to inference since given a set of random draws from the posterior distribution we can construct probability intervals simply by sorting the simulated values. For example, suppose we want an interval for the inverse of the mean of a normal distribution that contains this inverse mean with posterior probability $1 - \alpha$ (such an interval is called a probability interval or a credible set). If we have simulations of the mean we can invert all these simulated values, sort these and construct a $(1 - \alpha)$ probability interval by finding the $\alpha/2$ and $(1 - \alpha/2)$ percentiles of the simulated values. This saves us from having to approximate the distribution of a random variable that is defined by inverting a random variable whose distribution is known. In some cases, one can just use numerical integration (which is available in most software packages, such as R). When there are only a few parameters, numerical integration is often feasible, but the contemporary approach to these problems usually involves simulation.

2.3 Maximum likelihood estimates

When the data are iid and the likelihood obeys a number of technical conditions that are usually met for commonly used models, for large sample sizes, the posterior distribution becomes centered around the posterior mode, denoted $\hat{\theta}$. One important, non-technical assumption made is that the number of parameters does not increase as the sample size increases (which the iid assumption implicitly excludes). Hence we can Taylor expand the posterior around the posterior mode (provided this mode isn't on the boundary of the allowed set of θ)

$$
\begin{aligned}
p(\theta|y) &= \exp\{\log p(\theta|y)\} \\
&= \exp\Big\{ \log p(\hat{\theta}|y) + (\theta - \hat{\theta}) \frac{d}{d\theta} \log p(\theta|y)|_{\theta=\hat{\theta}} \\
&\quad + \frac{1}{2}(\theta - \hat{\theta})^2 \frac{d^2}{d\theta^2} \log p(\theta|y)|_{\theta=\hat{\theta}} \Big\}
\end{aligned}
$$

but

$$
\frac{d}{d\theta} \log p(\theta|y)|_{\theta=\hat{\theta}} = 0
$$

since $\hat{\theta}$ is the mode, so that

$$
p(\theta|y) = p(\hat{\theta}|y) \exp\Big\{ -\frac{1}{2}(\theta - \hat{\theta})^2 \Big[-\frac{d^2}{d\theta^2} \log p(\theta|y)|_{\theta=\hat{\theta}} \Big] \Big\}.
$$

This expansion indicates that the posterior distribution is approximately normal with mean $\hat{\theta}$ and variance $(-\frac{d^2}{d\theta^2} \log p(\theta|y)|_{\theta=\hat{\theta}})^{-1}$. If $p(\theta) \propto 1$ this mean is known as the maximum likelihood estimate (MLE) and the variance is the inverse of a quantity called the *observed Fisher information*, and is often denoted $I(\hat{\theta})$. When we refer to a distribution which holds when the sample size is large, we speak of an *asymptotic distribution*. As the sample size becomes increasingly large, the prior becomes increasingly irrelevant, hence if we have lots of data we don't need to be so concerned about the specification of the prior distribution. Non-Bayesian statistics relies

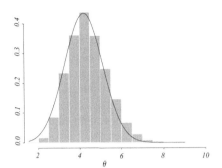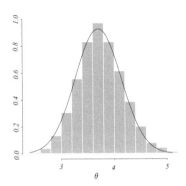

Figure 2.1 *Comparison of the posterior distribution (histogram) to the normal approximation (the density is represented as a line) for 2 sample sizes ($n = 5$ and $n = 20$).*

quite heavily on MLEs, and they use the same distribution we have shown is the approximate posterior, but since non-Bayesians typically think of the parameter fixed at some value and make statements about what will happen in repeated sampling from the data distribution, the interpretation of the results differ. An important frequentist result is that as the sample size increases, the MLE will get closer to the unknown value of θ. If an estimate has this property it is said to be *consistent*. Finally, although we have stressed the similarities between large sample arguments for frequentist approaches and Bayesian, we have argued that the posterior distribution converges pointwise to a normal density, whereas in the usual treatment of MLEs one only argues that the MLE converges in distribution to a normal density.

An important property of the MLE is that if we want the MLE of some monotone function of θ, say $\eta = f(\theta)$, and if $\hat{\theta}$ is the MLE of θ then the MLE of $f(\theta)$ is just $f(\hat{\theta})$. This follows from the fact MLEs are just numerical maximums of functions, hence provided we can invert f, we can recover the maximum using a different parameterization. Note that transformations lead to different posterior distributions since it is not true in general that $p(\eta)$ is just $p(f(\theta))$. In fact it is not hard to show that in this case

$$p(\eta) = \left| \frac{d\theta}{d\eta} \right| p\left(f^{-1}(\eta) \right),$$

where p on the right of this expression is the density of θ, alternatively

$$p_\eta(x) = \left| \frac{df^{-1}}{dx} \right| p_\theta\left(f^{-1}(x) \right).$$

We now return to the Poisson example investigated previously. We can examine the asymptotic approximation in this case. From 2.1, note that

$$\log p(\theta | y_1, \ldots, y_n) = \log C(y_1, \ldots, y_n) - n\theta + \left(\sum_{i=1}^{n} y_i \right) \log \theta,$$

so that if we take the first derivative of this expression, set the result equal to zero and solve the resulting equation for θ, we find that the posterior mode is $\bar{y}$. The second derivative is $-n\frac{\bar{y}}{\theta^2}$, so if we substitute the posterior mode for θ we get $-\frac{n}{\bar{y}}$ which gives an approximate variance of $\frac{\bar{y}}{n}$. Then using the approxiation that the posterior is approximately normal with mean given by the posterior mode and variance given by the negative inverse of the second derivative evaluated at the posterior mode, we find that our posterior is approximately $N(\bar{y}, \frac{\bar{y}}{n})$. Note that for large n our expression for the posterior mean, i.e., $\bar{y} + \frac{1}{n}$ is very close to the posterior mode and the posterior variance, i.e. $\frac{\bar{y}}{n} + \frac{1}{n^2}$ is even closer to the approximate posterior variance. We can also compare the exact posterior distribution to the approximate in this case. Figure 2.1 shows the approximate posterior densities along with histograms of samples from the posterior distribution for varying sample sizes (the figure used $n = 5$ and $n = 20$). Note that even for the $n = 5$ case the normal approximation is close to the true posterior distribution, and the $n = 20$ case indicates that the approximation is very close with this sample size. Finally, note here that the frequentist normal approximation coincides with our Bayesian approximation (due to our choice of prior distribution) so that the usual 95% confidence interval is such that the posterior probability that θ falls in this interval is 0.95.

As another example, suppose that y_i are independent for $i = 1, \ldots, n$ and $y_i \sim N(\mu, \sigma^2)$. Now

$$p(y_i|\mu, \sigma^2) = (2\pi\sigma^2)^{-\frac{1}{2}} \exp\left\{-\frac{(y_i - \mu)^2}{2\sigma^2}\right\},$$

so that the log-likelihood is

$$\ell(y_1, \ldots, y_n|\mu, \sigma) = -\frac{n}{2}\log(2\pi) - n\log\sigma - \frac{1}{2}\sum_{i=1}^{n}(y_i - \mu)^2.$$

If we differentiate this expression with respect to μ, substitute $\hat{\mu}$ for μ and set the result to 0 we find that $\hat{\mu} = \frac{1}{n}\sum_{i=1}^{n} y_i$. In a similar fashion we can find the MLE of σ^2, which we denote $\hat{\sigma}^2$ must satisfy the equation

$$\hat{\sigma}^2 = \frac{1}{n}\sum_{i=1}^{n}(y_i - \mu)^2,$$

so that upon substituting the MLE of μ we find

$$\hat{\sigma}^2 = \frac{1}{n}\sum_{i=1}^{n}(y_i - \bar{y})^2.$$

Hence if we have iid data from a normal distribution, the sample mean is the MLE of μ and the sample variance times $n/(n-1)$ is the MLE of σ^2.

Linear regression is a common extension of the normal example just presented. In the linear regression model, rather than assuming $y_i \sim N(\mu, \sigma^2)$, we have the model $y_i \sim N(\mu_i, \sigma^2)$ where $\mu_i = x_i'\beta$ for a given p-vector x_i specific to case i and a p-vector of regression coefficients β. We still assume that the y_i are independent for

$i = 1, \ldots, n$ but they no longer have the same mean. In this case, the log-likelihood is

$$\ell(y_1, \ldots, y_n | \beta, \sigma) = -\frac{n}{2} \log(2\pi) - n \log \sigma - \frac{1}{2} \sum_{i=1}^{n} (y_i - x_i'\beta)^2.$$

To find the MLE of β we differentiate $\ell(y_1, \ldots, y_n | \beta, \sigma)$ with respect to β to get

$$\frac{\partial \ell}{\partial \beta} = \sum_{i=1}^{n} 2(y_i - x_i'\beta) \frac{\partial}{\partial \beta}(-x_i'\beta).$$

Differentiating a scalar with respect to a p-vector gives a p-vector with i^{th} component given by the derivative of the scalar with respect to the ith element of the p-vector that one is differentiating with respect to. This implies

$$\frac{\partial}{\partial \beta}(-x_i'\beta) = -x_i$$

so that

$$\frac{\partial \ell}{\partial \beta} = -\sum_{i=1}^{n} 2x_i(y_i - x_i'\beta).$$

If we set this equal to zero and solve for β we find

$$\sum_{i=1}^{n} x_i y_i = \sum_{i=1}^{n} x_i x_i' \hat{\beta}$$

or, noting that $\sum_i x_i x_i'$ is a $p \times p$ matrix

$$\hat{\beta} = (\sum_{i=1}^{n} x_i x_i')^{-1} \sum_{i=1}^{n} x_i y_i.$$

Frequently this expression is more compactly expressed using linear algebra. This is done by letting y by a vector with i^{th} element y_i and letting X denote the $n \times p$ matrix with i^{th} row x_i. We then find that

$$\hat{\beta} = (X'X)^{-1}X'y.$$

We also find that this is the estimate of β that makes $\sum_{i=1}^{n}(y_i - x_i'\beta)^2$ minimal, an estimate known as the *ordinary least squares estimate*.

Finally we will consider the multinomial distribution. The multinomial distribution is a commonly used generalization of the binomial distribution. The binomial distribution is used to model the number of successes (denoted y) in a fixed number (say n) of independent trials with common success probability, denoted θ. Then the likelihood is given by the expression

$$L(y|\theta) = \binom{n}{y} \theta^y (1 - \theta)^{n-y}.$$

Suppose we set $z_i = (1, 0)$ if the ith trial results in a success and set $z_i = (0, 1)$ if the ith trial is a failure. If we then set $x = \sum_i z_i$ and denote the ith element of x as

x_i, then $x_1 = y$ and $x_2 = n - y$). If we then set $\pi_1 = \theta$ and $\pi_2 = 1 - \theta$ then we can write the likelihood for x as

$$L(x|\pi_1, \pi_2) = \left(\frac{(x_1 + x_2)!}{x_1! x_2!}\right) \pi_1^{x_1} \pi_2^{x_2}$$

where we require $\pi_1 + \pi_2 = 1$. Now suppose there are m possible outcomes rather than just 2. In this case we define z_i to be an m-vector with components 0 and 1 where the kth component of z_i is 1 if z_i is in category k and 0 otherwise. We can then easily generalize our previous expression for the likelihood of $x = \sum_{i=1}^{m} z_i$ as

$$L(x|\pi_1, \ldots \pi_m) = \left(\frac{(\sum_{i=1}^{m} x_i)!}{\prod_{i=1}^{m} x_i!}\right) \prod_{i=1}^{m} \pi_i^{x_i},$$

where we assume $\sum_{i=1}^{m} \pi_i = 1$.

Bayesian analysis of the multinomial model requires specification of a prior distribution on the set of π_i and requires $\sum_i \pi_i = 1$. For computational ease, a *conjugate prior* is typically specified for this collection of parameters. A conjugate prior is a prior distribution that, when combined with the likelihood, results in a posterior distribution with the same form as the prior. Such priors are frequently used to simplify computation since the posterior distribution will have a fixed, known form that we can usually draw samples from using common methods available in most software. The conjugate prior for the multinomial likelihood is a generalization of the conjugate prior used in the analysis of the binomial likelihood, hence we first cover the binomial case.

The conjugate prior for the binomial likelihood is given by the *beta distribution* with density

$$\frac{\Gamma(\alpha + \beta)}{\Gamma(\alpha)\Gamma(\beta)} \theta^{\alpha-1} (1 - \theta)^{\beta-1}$$

for 2 parameters α and β. First note that if we multiply the binomial likelihood by this prior then the result has the same form as the prior, i.e. the posterior is also a beta distribution, but now with parameters $y + \alpha$ and $n - y + \beta$. Next, note that the prior mean is $E\theta = \alpha/(\alpha + \beta)$ and the variance is $E[\theta](1 - E[\theta])/(\alpha + \beta + 1)$ so that the relative sizes of α and β determine the prior mean and the sum of α and β determine the prior variance. Hence a minimally informative prior can be obtained by making α and β small (they are also usually taken to be equal so that the prior mean is $\frac{1}{2}$). We can generalize this by using the previously introduced notation for π_i and by setting $\alpha_1 = \alpha$ and $\alpha_2 = \beta$. The beta distribution now has the form

$$\frac{\Gamma(\sum_{i=1}^{2} \alpha_i)}{\prod_{i=1}^{2} \Gamma(\alpha_i)} \prod_{i=1}^{2} \pi_i^{\alpha_i - 1}.$$

If we now suppose there are m possible outcomes then we can use the same specification of the prior, although it is now called the *Dirichlet distribution* with parameter vector $\alpha = (\alpha_1, \ldots, \alpha_m)$. While it is not obvious that this expression integrates to 1 over the set of $\pi_i > 0$ so that $\sum_i \pi_i = 1$, this is indeed the case. Components of the parameter vector α have a similar interpretation as in the Beta case. In particular, the prior mean for the ith component of α is $\alpha_i / \sum_i \alpha_i$ and the prior variance is inversely proportional to $\sum_i \alpha_i$. Finally the posterior distribution of θ is Dirichlet with parameter $\alpha + x$.

2.3.1 The EM algorithm

The EM algorithm is a method that is commonly used to find posterior modes and MLEs. It is well suited to this task for many commonly used statistical models, and will be discussed further at several points later. We will suppose that we are using the algorithm to find a posterior mode, but the application to finding MLEs is almost identical. Suppose that we can partition the parameters, into 2 sets, which we denote θ and η and our goal is to find the marginal posterior mode of θ, that is, we seek to maximize $p(\theta|y)$. For example, θ might represent the parameters in a statistical model and η might be missing values for some of variables. Another example is that η is a vector of indicator variables that indicate the category of each subject where there are 2 possible categories, but category membership is not observed. In such a case, we might not really be concerned with how subjects get classified, but we do care about parameters related to the marginal distribution of the parameters in the model.

Central to the algorithm is a function that is called the expected value of the complete data log-posterior (or log-likelihood), and is denoted $Q(\theta|\theta_i) = E_i \log p(\theta, \eta|y)$. In this expression, the expectation only treats the parameters η as random and everything else is treated as a constant, and the distribution of η is computed given a value of θ_i. Computing $Q(\theta|\theta_i)$ is often called the E-step. It can be difficult to obtain this expression since one must find the expected value of η conditional on $\theta = \theta_i$, but for many models this is straightforward. After the E-step, we then maximize $Q(\theta|\theta_i)$ with respect to θ, which is referred to as the M-step. If we denote the maximizing value of $Q(\theta|\theta_i)$ by θ_{i+1}, we thereby generate a sequence of θ_i values that will converge to a local posterior mode. If there are multiple local modes, the selection of the starting value θ_1 can impact the final estimate. For this reason, it is usually a good idea to try several sets of reasonable initial values based on the data. Frequently such initial estimates can be obtained by ignoring some of the data (such as cases with some missing data) or using simple estimates.

As an example, suppose we have $n_1 + n_2$ iid Bernoulli trials but can't determine if n_2 of these trials were successes or failures. For the n_1 trials whose outcome we can determine, suppose there were y successes. We use θ to represent the probability of a success and for convenience we assume that the prior is uniform on the interval $(0,1)$. Let η represent the number of successes among the n_2 trials (which is unobserved). Then we can write an expression for the expected value of the complete data log-posterior on the ith iterations by observing that $y + \eta \sim \text{Bin}(n_1 + n_2, \theta)$, so that

$$
\begin{aligned}
Q(\theta|\theta_i) &= \text{E}_i\left[\log\binom{n_1 + n_2}{y + \eta}\theta^{y+\eta}(1 - \theta)^{n_1+n_2-y-\eta}\right]\\
&= \text{E}_i\left[\log\binom{n_1 + n_2}{y + \eta}\right] + \text{E}_i\left[(y + \eta)\log\theta\right]\\
&\quad + \text{E}_i\left[(n_1 + n_2 - y - \eta)\log(1 - \theta)\right].
\end{aligned}
$$

Recall that in the E-step we compute the expected value of the missing data (here η) given the current value of the observed data and the parameters (here y and θ_i). This implies that θ and y pass through the expectation so that we obtain

$$
\begin{aligned}
Q(\theta|\theta_i) &= \text{E}_i\left[\log\binom{n_1 + n_2}{y + \eta}\right] + (y + \text{E}_i[\eta])\log\theta\\
&\quad + (n_1 + n_2 - y - \text{E}_i[\eta])\log(1 - \theta).
\end{aligned}
$$

While the first term in this expression cannot be simplified, this poses no problem since when we maximize this expression with respect to θ it plays no role since this term is not a function of θ (note that it will be a function of θ_i but that is fine). Such terms frequently arise when using the EM algorithm. Next, note that $\eta|\theta_i \sim \text{Bin}(n_2, \theta_i)$ so that $\text{E}_i[\eta] = n_2\theta_i$. Hence

$$
\begin{aligned}
Q(\theta|\theta_i) &= \text{E}_i\left[\log\binom{n_1 + n_2}{y + \eta}\right] + (y + n_2\theta_i)\log\theta\\
&\quad + (n_1 + n_2 - y - n_2\theta_i)\log(1 - \theta).
\end{aligned}
$$

The M-step then consists of maximizing this expression with respect to θ. Frequently this maximization must be done numerically, but for this simple example, we can optimize this expression by differentiating $Q(\theta|\theta_i)$ with respect to θ, setting the result to zero and solving for θ to obtain an expression for θ_{i+1} that depends on y and θ_i. Differentiating

$$
\frac{dQ}{d\theta} = \frac{y + n_2\theta_i}{\theta} - \frac{n_1 + n_2 - y - n_2\theta_i}{1 - \theta}
$$

then setting to zero, substituting θ_{i+1} for θ and solving for θ_{i+1} gives

$$
\theta_{i+1} = \frac{y + n_2\theta_i}{n_1 + n_2}.
$$

In typical uses one must use a computer to recursively solve this equation given some θ_1, and run sufficient numbers of these recursions so that the value of θ_i stabilizes. Here, however, we can solve this system explicitly. To do so, take the limit on both sides of the equation, and use the fact that $\lim_{i\to\infty}\theta_i = \lim_{i\to\infty}\theta_{i+1}$ here to obtain an expression for $\lim_{i\to\infty}\theta_i$. This results in $\lim_{i\to\infty}\theta_i = \frac{y}{n_1}$, an estimate which is the MLE if we just ignored the the trials whose outcome could not be determined.

2.4 Likelihood ratio tests

Hypothesis testing can be thought of as trying to use the data to decide which of 2 competing models, here denoted M_1 and M_2, is more likely. We will here discuss the special situation in which M_2 is a special case of M_1. As an example, suppose M_1 specifies that $y_i \sim N(\mu, 1)$ and M_2 specifies $y_i \sim N(0, 1)$, equivalently, we specify the null hypothesis that $\mu = 0$ and want to test this against the 2 sided alternative. Since models often depend on parameters, as in the previous example, it makes sense to treat these models as random objects and enquire about the posterior probability of a model $p(M_1|y)$. If we want to compare 2 models then we should look at the ratio of the posterior probabilities. Using Bayes theorem it is easy to show

$$\frac{p(M_1|y)}{p(M_2|y)} = \frac{p(y|M_1)}{p(y|M_2)} \frac{p(M_1)}{p(M_2)}.$$

This states that the posterior odds $\frac{p(M_1|y)}{p(M_2|y)}$ is equal to the product of the prior odds, $\frac{p(M_1)}{p(M_2)}$, and the likelihood ratio, $\frac{p(y|M_1)}{p(y|M_2)}$. If we suppose that prior to seeing the data the 2 models are equally likely, then the prior odds is one so the posterior odds is determined solely by the likelihood ratio. Likelihood ratios provide a way of testing a hypothesis in a general probability model since one only needs to be able to write down the likelihood in order to compute the likelihood ratio.

In order to calculate the p-value for a hypothesis based on the likelihood ratio, we need to determine the distribution of the likelihood ratio assuming the null hypothesis is true. It transpires that under assumptions similar to those used to prove that the posterior is asymptotically normally distributed, $2 \log \frac{p(y|M_1)}{p(y|M_2)}$ is distributed according to a χ^2 distribution with degrees of freedom given by the difference in the number of parameters in the 2 models. Note that the p-value we obtain in this fashion is not a posterior probability but rather is the usual sort of frequentist p-value.

2.4.1 Maximized likelihood ratio tests

Frequently hypotheses are formulated about only a subset of the parameters in the model. For example, suppose y_i are iid random variables which are assumed to be $N(\mu, \sigma^2)$ and we wish to test the hypothesis that $\mu = 0$, In this case we don't care about the variance σ^2. One can still use likelihood ratio tests to compute p-values for hypothesis tests, but since the ratio may depend on the parameters that are not restricted by the null hypothesis, one must substitute the maximum likelihood estimates of these parameters. For the normal example, just mentioned, $2 \log$ of the likelihood ratio is just $\frac{\bar{y}^2}{\sigma^2/n}$, so that σ appears in the likelihood ratio test (see exercise 4). If σ was known, then since the distribution of the sum of a set of normally distributed random variables is also normally distributed, we find that under the null hypothesis, $\bar{y} \sim N(0, \sigma^2/n)$, so that twice the log of the likelihood ratio, here $\frac{\bar{y}^2}{\sigma^2/n}$, will indeed be distributed like a χ^2 random variable with 1 degree of freedom (since it is the square of a $N(0, 1)$ random variable). If σ^2 is unknown, then we can find its MLE, namely $\frac{1}{n} \sum_i (y_i - \bar{y})^2$ and substitute this for the unknown value of σ^2. For large sample sizes this will not affect the distribution of the likelihood ratio test since we will have an accurate estimate of σ^2.

2.5 Empirical Bayes analysis

Another approach to likelihood based statistical inference is known as empirical Bayes. This approach estimates parameters that appear in the prior distribution using the observed data, then uses these estimates in the prior distribution as if they were known prior to seeing the data. The estimates for the parameters in the prior are usually obtained via maximum likelihood, but other methods (such as the method of moments) are sometimes used. As substituting data based estimates into the prior distribution invalidates the basis for Bayes theorem, such methods are usually justified in terms of their frequentist properties (i.e. mean squared error). As an example, suppose that y_i for $i = 1, \ldots, n$ is distributed according to a binomial distribution so that $y_i \sim \text{Bin}(m_i, \theta_i)$ where the y_i given θ_i are assumed independent. Further suppose the set of θ_i are iid from a beta distribution with parameters α and β. We then obtain a likelihood for the data in terms of the parameters α and β. For this case we have

$$
\begin{aligned}
L(y_1, \ldots, y_n | \alpha, \beta) &= \int_0^1 \cdots \int_0^1 p(y_1, \ldots, y_n, \theta_1, \ldots \theta_n | \alpha, \beta) \, d\theta_1 \cdots d\theta_n \\
&= \int_0^1 \cdots \int_0^1 p(y_1, \ldots, y_n | \theta_1 \ldots \theta_n, \alpha, \beta) \\
&\quad \times \; p(\theta_1, \ldots \theta_n | \alpha, \beta) \, d\theta_1 \cdots d\theta_n \\
&= \int_0^1 \cdots \int_0^1 \left[\prod_{i=1}^n p(y_i | \theta_i) \right] \left[\prod_{i=1}^n p(\theta_i | \alpha, \beta) \right] d\theta_1 \cdots d\theta_n \\
&= \int_0^1 \cdots \int_0^1 \prod_{i=1}^n \binom{m_i}{y_i} \theta_i^{y_i} (1 - \theta_i)^{m_i - y_i} \frac{\Gamma(\alpha + \beta)}{\Gamma(\alpha)\Gamma(\beta)} \\
&\quad \times \; \theta_i^{\alpha-1} (1 - \theta_i)^{\beta-1} \, d\theta_1 \cdots d\theta_n \\
&= \prod_{i=1}^n \binom{m_i}{y_i} \frac{\Gamma(\alpha + \beta)}{\Gamma(\alpha)\Gamma(\beta)} \int_0^1 \theta_i^{y_i + \alpha - 1} (1 - \theta_i)^{m_i - y_i + \beta - 1} \, d\theta_i \\
&= \prod_{i=1}^n \binom{m_i}{y_i} \frac{\Gamma(\alpha + \beta)}{\Gamma(\alpha)\Gamma(\beta)} \frac{\Gamma(y_i + \alpha)\Gamma(m_i - y_i + \beta)}{\Gamma(\alpha + \beta + m_i)}
\end{aligned}
$$

a derivation that follows from the structure of the model and the fact that the beta distribution is a density (and so integrates to 1). One could then optimize this function using a numerical optimization procedure to obtain the MLEs of α and β. An empirical Bayes approach to this problem would then use these MLEs, denoted $\hat{\alpha}$ and $\hat{\beta}$ to conduct inference for the individual θ_i. For example, for this problem, the posterior distribution of θ_i given α and β is $\text{Beta}(y_i + \alpha, m_i - y_i + \beta)$ and this distribution has mean $\frac{y_i + \alpha}{m_i + \alpha + \beta}$, so that the simple empirical Bayes estimate of θ_i is just $\frac{y_i + \hat{\alpha}}{m_i + \hat{\alpha} + \hat{\beta}}$.

Intuitively, the problem with this approach is that one is using the data to estimate the parameters in the prior and then treats these values as known when one then uses the Bayes approach. There have been a number of attempts to correct for this effect (e.g. make the reported 95% probability intervals larger by some amount), and while these can be useful, most researchers would just use a fully Bayesian approach to the estimation of α and β (i.e. put priors on these parameters and treat them like other parameters in a statistical model). However note that as n increases for the previous example, we will accumulate more information about α and β hence uncertainty about these parameters becomes less relevant if we are primarily concerned with the set of θ_i.

More generally, suppose one has a model that specifies $p(y|\theta)$ and a prior for θ that depends on a collection of parameters ϕ. Moreover, suppose the model and data structure are so that the posterior distribution of ϕ can be approximated by a *Dirac delta function* (or a point mass) at $\hat{\phi}$. For most practical purposes we can just think of Dirac delta function as $\lim_{k \to 0} 1_{\{x-k, x+k\}}/(2k)$. This approximation would be reasonable if the data are highly informative about this vector of parameters, and $\hat{\phi}$ is the posterior mode. Then

$$
\begin{aligned}
p(\theta|y) &= \int p(\theta, \phi|y) \, d\phi \\
&= \int p(\theta|y, \phi) p(\phi|y) \, d\phi \\
&\approx p(\theta|y, \hat{\phi}).
\end{aligned}
$$

This suggests that one can approximate the marginal posterior distribution of θ by just substituting in the posterior mode for the parameters that appear in the prior of θ. This approximation can simplify computations since one needn't integrate ϕ out of its joint posterior with θ. To determine $\hat{\phi}$, one typically proceeds by finding the marginal posterior of ϕ and then maximizing it using some optimization algorithm. Many models used in genetics and genomics have this structure, thus empirical Bayes methods can be useful approximations to the fully Bayesian approach.

2.6 Markov chain Monte Carlo sampling

Often the posterior of θ is not approximately normal and not of any standard form, hence other methods are necessary to obtain simulations from this distribution. While a number of specialized approaches have been developed for specific problems, one very general way to draw samples from a probability distribution is to use the *Metropolis algorithm*. This algorithm proceeds by constructing a Markov chain whose limiting distribution is the distribution from which you want samples. Such methods are referred to as Markov chain Monte Carlo methods (MCMC). To use this class of algorithms, one simulates a long draw from the Markov chain, then uses samples from the chain once it has converged to its limiting distribution. In practice, determining whether the chain has converged to its limiting distribution is difficult. One sensible approach is to run several independent chains and examine how long it takes for them to mix. Plots of the sampled value of the parameter plotted against the number of iterations usually reveal when the chains have converged to the same area of parameter space. In addition, there are a number of diagnostic measures that can be used to assess if a collection of independent runs from a Markov chain has converged to its limiting distribution (see, for example, Gelman and Rubin, 1992).

But how can one construct a Markov chain whose limiting distribution is $p(\theta|y)$? This turns out to be surprisingly simple. Suppose we have some distribution over the sample space from which we can draw simulations given the current value of the chain. Call this distribution the jumping distribution, and denote it $J(\theta|\theta_t)$. We require that jumping distributions are symmetric in the sense that $J(\theta_a|\theta_b) = J(\theta_b|\theta_a)$, but an extension of the Metropolis algorithm called the Metropolis-Hastings algorithm allows one to use non-symmetric jumping distributions. For example we could use a normal density with mean given by θ_t and variance given by the negative inverse of the matrix of second derivatives of the log posterior distribution (an approach that often works quite well). Normal densities clearly have the desired symmetry and are easy to sample from. We need to choose some initial value for our chain, denote it θ_0. The algorithm proceeds by drawing a simulation from the jumping distribution, call it θ^*, then evaluating the ratio, r, of the posterior at θ^* to the posterior at the current location of the chain, θ_{t-1}

$$r = \frac{p(\theta^*|y)}{p(\theta_{t-1}|y)}.$$

Then $\theta_t = \theta^*$ with probability $\min(r, 1)$ otherwise $\theta_t = \theta_{t-1}$. Notice that the value of θ_t will not change with each iteration. The proportion of jumps that are accepted indicates how effectively the algorithm is working. Generally one wants to find that about 20-40% of the jumps are accepted. If the rate of accepted moves is not in this range, then one should multiply the covariance matrix of the jumping distribution by some constant so that the acceptance rate is in this range. Note that by making the covariance smaller, the proposed moves will be close to the current location of the chain and so r, on average, will increase thereby increasing the accepted proportion of moves. Conversely, making the covariance large will tend to make the accepted proportion of moves decrease. Finally, note that

$$
\begin{aligned}
r &= \frac{p(\theta^*|y)}{p(\theta_{t-1}|y)} \\
&= \frac{L(y|\theta^*)p(\theta)/p(y)}{L(y|\theta_{t-1})p(\theta_{t-1})/p(y)} \\
&= \frac{L(y|\theta^*)p(\theta)}{L(y|\theta_{t-1})p(\theta_{t-1})}
\end{aligned}
$$

so that we don't need an expression for $p(y)$ at all to use this algorithm. This is good since we frequently cannot find a closed form expression for $p(y)$ since we cannot analytically compute the integral $\int L(y|\theta)p(\theta)\,d\theta$.

To see that this algorithm will indeed draw samples from the limiting distribution of the chain, first suppose that the state space and jumping distributions are such that the chain is aperiodic and irreducible. These 2 conditions are typically met in most applications. Then there exists a limiting distribution for the chain. We now show that this limiting distribution is the posterior distribution we seek. To this end, consider the joint distribution of 2 adjacent values in the chain, $p(\theta_t, \theta_{t-1})$ (implicitly conditioning on the data). Using $p(y)$ to denote the posterior density at y, suppose $p(x) > p(y)$. Then

$$p(\theta_t = x, \theta_{t-1} = y) = p(y)J(x|y),$$

and also

$$p(\theta_t = y, \theta_{t-1} = x) = p(x)J(y|x)r,$$

where $r = \frac{p(y)}{p(x)}$. But then

$$
\begin{aligned}
p(\theta_t = y, \theta_{t-1} = x) &= p(y)J(y|x) \\
&= p(y)J(x|y),
\end{aligned}
$$

by symmetry of the jumping distribution. But then

$$
\begin{aligned}
p(\theta_t = x) &= \sum_y p(\theta_t = x, \theta_{t-1} = y) \\
&= \sum_y p(\theta_t = y, \theta_{t-1} = x) \\
&= p(\theta_{t-1} = x).
\end{aligned}
$$

This implies that the marginal distributions at each time point are the same. Therefore if θ_{t-1} is a draw from the posterior distribution, then so is θ_t. Note that these will not be independent draws (it is a Markov chain).

As an alternative to the Metropolis algorithm, we can use the *Gibbs sampler* to draw samples from the posterior distribution. To illustrate the idea of this method, suppose θ is a vector and denote the ith element of θ as θ_i. The algorithm proceeds by finding what is known as the full conditional distribution for each parameter, which is just $p(\theta_i|y, \theta_{j\neq i})$, then one draws a sample from this distribution (for many statistical models these full conditional distributions have simple closed form expressions). Call the sampled value on the tth iteration for element θ_i $\theta_i^{(t)}$. One then uses this value when sampling θ_j from its full conditional for $j > i$. For example, if θ is a 3 vector and $\theta_i^{(0)}$ are a set of initial values, we first sample from the distribution

$$p(\theta_1|y, \theta_2^{(0)}, \theta_3^{(0)})$$

to get $\theta_1^{(1)}$, then sample from

$$p(\theta_2|y, \theta_1^{(1)}, \theta_3^{(0)})$$

to get $\theta_2^{(1)}$, then sample from

$$p(\theta_3|y, \theta_1^{(1)}, \theta_3^{(1)})$$

to get $\theta_3^{(1)}$. This algorithm also generates a Markov chain whose limiting distribution is the posterior distribution (and so is an MCMC algorithm). To obtain an expression for $p(\theta_i|y, \theta_{j\neq i})$, one first obtains an expression for the joint posterior distribution, $p(\theta|y)$. Then we obtain an expression proportional to $p(\theta_i|y, \theta_{j\neq i})$ by just considering those factors that involve θ_i.

2.7 Bibliographic notes and further reading

As with the previous chapter, there are a large number of resources to learn more about the material discussed here. As for texts with a Bayesian focus, 2 excellent sources are Gelman et al., (2004) and Carlin and Louis (2009). A succinct and complete treatment of frequentist statistics can be found in Silvey (1975), and a more text like, and perhaps easier to understand version of the same material can be found in Rice (1995).

2.8 Exercises

1. Suppose y is binomially distributed, i.e. $P(Y = k) = \binom{n}{k}\theta^k(1-\theta)^{n-k}$. Compute the MLE of θ and find its asymptotic variance.

2. If y is a vector giving the frequencies with which each of the categories of a multinomial observation where observed, show that the posterior mean of the ith component of θ is given by the sample proportion of this category if $\alpha_i = 0$ for all i (assuming that there is an observation from each of the possible categories).

3. Show that the posterior odds is equal to the product of the likelihood ratio and the prior odds.

4. Verify the claim made in section 2.4.1 that $2 \log$ of the likelihood ratio is just $\frac{\bar{y}^2}{\sigma^2/n}$.

5. Suppose that $y_i \sim N(0, \sigma^2)$ for $i = 1, \ldots, n$ and that all of the y_i are independent. Suppose that σ^{-2} has a gamma prior for some pair of parameters α and β. What is the posterior distribution of σ^{-2}? Also find the posterior mean and variance.

6. Use the proof that the limiting distribution of the Metropolis algorithm is the posterior distribution to determine the appropriate choice of r for the Metropolis-Hastings algorithm. This is the only difference between the Metropolis algorithm and the Metropolis-Hastings algorithm.

7. Suppose y_i are iid observations from a 2-component normal mixture model with density

$$\lambda(2\pi\sigma^2)^{-\frac{1}{2}}\exp\left\{-\frac{(y-\mu_1)^2}{2\sigma^2}\right\} + (1-\lambda)(2\pi\sigma^2)^{-\frac{1}{2}}\exp\left\{-\frac{(y-\mu_2)^2}{2\sigma^2}\right\}$$

(So both components have the same variance). Use flat and independent priors for λ, $\log \sigma$, μ_1 and μ_2. Let η_i be an indicator variable that indicates which component of the mixture model observation i comes from (these indicator variables are unobserved). Suppose that the η_i are Bernoulli random variables. Treating the η_i as missing data, show how to use the EM algorithm to find the posterior mode of λ, $\log \sigma$, μ_1 and μ_2.

8. Suppose that spheres of identical size have centers which are distributed uniformly on the 3 dimensional cube $[0, 1]^3$. Let (x, y, z) represent a point in this unit cube. We take a cross section of the cube at height $z = \frac{1}{2}$. This cross-section has n circles, and these circles have radii $y_1, \ldots, y_n$. You may assume that the radii of these spheres, which we denote θ, is less than 0.5.

   (a) What is the maximum likelihood estimate (MLE) of the volume of these spheres?

   (b) Define the random variable $Y = n\sqrt{1 - \left(\frac{\hat{\theta}}{\theta}\right)^2}$, where $\hat{\theta}$ is the MLE of θ. Find the distribution function of Y.

   (c) Take the limit as $n \to \infty$ in your expression for the distribution function of Y from part B. Is this asymptotic distribution what you would expect based on the usual theory of maximum likelihood? If not explain why.

   (d) Use the result form part C to find an approximate $(1 - \alpha)\%$ confidence interval for θ of minimal width.

9. Let y be an n-vector and suppose $y \sim N(X\beta, \Sigma)$, where Σ is known and positive definite, X is a known matrix of full rank and β is a p-vector ($p < n$). Suppose we specify a prior for β that is $N(m, A)$ for some vector m and some positive definite matrix A (both known).

   (a) Derive the posterior distribution of β conditional on y and all known quantities.

   (b) Suppose $p = 2$. Provide details regarding how one could use the Gibbs sampler to sample from the joint bivariate distribution by sampling each of the univariate conditional distributions. We will use the notation $\beta_i^{(t)}$ to represent the simulated value of the ith element of β on the tth iteration of the algorithm.

   (c) Let $\bar{\beta}_1^{(n)} = \frac{1}{n} \sum_{j=1}^{n} \beta_1^{(j)}$ be a sample mean computed from samples generated by the Gibbs sampler from the previous part. Show that $E\bar{\beta}_1^{(n)} = b_1 + g(n)$ where $\lim_{n \to \infty} n^\alpha g(n) = 0$ for all $\alpha < 1$ for any choice of $\beta_2^{(0)}$ where b_1 is the posterior mean for the first element of β found in part A and the expectation is with respect to the probability measure generated by the Gibbs sampler from part B.

   (d) Show directly that samples generated according to the Gibbs sampler from part B converge in distribution to the posterior distribution from part A given any choice for the initial values.

10. Suppose that $x \sim \text{Bin}(n, \theta)$ and we observe $y = \max(x, n - x)$. Assume that n is an even number.

   (a) Give an expression for $p(\theta|y)$ using a uniform prior for θ.

   (b) Show that $p\left(|\theta - \frac{1}{2}| \leq C|x\right) = p\left(|\theta - \frac{1}{2}| \leq C|y\right)$ for all $C \in \left(0, \frac{1}{2}\right)$ using a uniform prior for θ.

(c) Using a uniform prior for θ, show that there will be a single posterior mode when

$$\frac{1}{2} \leq \frac{y}{n} \leq \frac{1}{2} + \frac{1}{2\sqrt{n}}.$$

Determine the value of the single posterior mode in this case.

(d) Show that when the previous inequality fails to hold there will be 2 posterior modes, θ_1 and θ_2, so that $p(\theta_1|y) = p(\theta_2|y)$. Indicate why this implies that the theory of maximum likclihood is not applicable in this case.

11. Suppose that $y_i \sim N(\theta_i, \sigma^2)$ for $i = 1, \ldots, n$, all the y_i are independent and σ^2 is known. If we suppose that θ_i has a $N(\mu, \tau^2)$ prior for all i, find the empirical Bayes estimate of θ_i. Describe when this estimate will not be in the interval bounded y_i and $\bar{y}$.

CHAPTER 3

Markers and Physical Mapping

3.1 Introduction

There have been tremendous advances in the mapping of genomes, the most notable being the recent completion of the human genome project (i.e. the sequencing of the human genome). The completion of these projects allows for isolation and manipulation of individual genes, and by cataloguing all genes in an organism we can start to take new approaches to answering some of the mysteries surrounding genomes, such as the function of non-coding DNA and the co-regulation of the use of genes. In addition, by knowing how genes work together to give rise to observed phenotypes, we can potentially produce medicines that alter protein expression in specific tissues so as to treat disease. Finally, if we know the location of disease genes, we can advise potential parents with regard to risk for disease or attempt to develop treatments based on knowledge of the gene products. Knowledge of the location of the genes along the genome is knowledge of the *physical map* of the organism. A *genetic map* relates how close 2 loci on a genome are in terms of risk for a certain trait based on knowledge of the allele at the other locus. Either sort of map provides information about the locations of genes.

To determine the sequence of nucleotides present in a DNA molecule, one starts with a large number of copies of a single stranded DNA molecule. These DNA molecules are stored in *clones* which are either bacteria, yeast or rodent cells that contain a segment of human DNA, and the cell that holds the sequence is referred to as a vector. A clone library is a collection of clones that cover a genome or chromosome. One then florescence labels a collection of nucleotides using 4 different florophores so that the identity of a nucleotide can be determined by a machine designed to detect each of the florophores used. In addition to this collection of labeled nucleotides, one creates a collection of molecules similar to the 4 nucleotides called dideoxynucleotides (ddNTPs) that differ from normal nucleotides in that as soon as a ddNTP is incorporated in a sequence of nucleotides, the sequence of nucleotides is terminated (there is a ddNTP that corresponds to each of the 4 usual nucleotides). One then combines the single stranded DNA molecules, the labeled nucleotides and the ddNTPs with compounds that promote copying the single stranded DNA molecule. When this happens, chains of labeled nucleotides are built up that terminate whenever a ddNTP is incorporated into the sequence. The result is a set of nucleotide sequences that differ in length and are copies of a portion of the original single stranded DNA sequences. The lengths of the fragments can be controlled by varying the concentration of the ddNTPs. The fragments are then separated using a denaturing polyacrylamide gel (i.e. a viscous material that has a denaturing compound embedded in it so that the DNA stays in its single stranded form). The larger fragments will not move as far in the gel as the lighter fragments, hence the set of fragments can be separated. One can then use a machine that reads the florophores embedded in the gel to determine the sequence of the starting DNA sequence. This process has been further automated using capillary-based DNA sequencing in which thin tubes are used in place of gels. A problem with this approach is that one can only sequence DNA fragments of length up to 750-1000 bases. To understand the extent that this is a limitation, note that the long arm of chromosome 22 (the shortest chromosome) is about 33.4 million basepairs (the long arm of chromosome 22 is the only part of that chromosome with DNA that is known to code for proteins).

For small genomes the *shotgun method* has provided a means of determining the sequence of a genome. This approach attempts to overcome the discrepancy between the length of a DNA molecule that can be sequenced and the typical size of a DNA molecule by first breaking up the molecule into smaller, overlapping segments, sequencing these, then putting all the fragments back together. As an example, suppose that an organism has the following DNA sequence on one strand

        AACATGTGTGTCAATATACGCGGCCTTTATATATATGTGCGTGCGTGC

Suppose one has a large collection of these sequences and then fragments them to obtain many subsequences, of which suppose 5 are

        AACATGTGTGTCAATATA
                GTCAATATACGCGGCC
                    TATACGCGGCCTTTAT
                        GGCCTTTATATATATGT
                            ATATATATGTGCGTG
                                TGTGCGTGCGTGC
                                    GTGCGTGC

If one didn't know the original sequence, but knew the fragmented sequences, one can deduce the original sequence. While the amount that 2 fragments must overlap in general depends on the size of the fragments and the length of the genome being sequenced, in practice roughly at least 40 nucleotides must overlap before a typical sequence assembly algorithm will combine the fragments (Lander and Waterman, 1988). Once a set of fragments has been combined, they are referred to as a *contig*. To assemble the sequenced fragments, one could label all of the fragments, then consider every possible permutation of these fragments and choose the shortest sequence that is consistent with all of the fragments. Optimizing a function over a set of permutations is a well studied problem, and is of the type generally referred to as the *traveling salesman problem*. While exact solutions to these sorts of problems can be obtained through enumeration, this is often takes too much time (since the number of possible permutations increases exponentially with the number of fragments).

As an alternative, approximate solutions can be obtained using randomized search procedures, such as simulated annealing (see Press et al., 1992).

Unfortunately, for complex organisms, putting all the segments back together correctly is very difficult since the sequences are very long relative to the size of the DNA sequences one can insert into a vector for use in a clone library (recall the human genome is about 3 billion bp, or 3,000 megabase pairs, Mb), and there are many subsequences that are repeated at different locations throughout the genome. The largest clones one can use for constructing a clone library for humans (namely, yeast artificial chromosomes) require thousands of clones, hence the number of possible permutations of the fragments is too large to consider every possible permutation. In addition, there is a lot of repetitive subsequences in DNA. For example, in the human genome there are repeats that are 100s of basepairs in length that occur thousands of times in the genome. Once a set of contigs has been constructed, it is possible to fill in the gaps by directed sequencing of the missing fragments, however, as this is difficult to automate (in contrast to random fragmentation and assembly) there is a trade off in terms of how many repeated fragmentations of the genome one uses for assembly and filling in the gaps.

One approach to deal with the problem of sequence assembly in complex organisms is to break up the genome into known short segments that can be sequenced, sequence these (using the shotgun method), then put all these known short segments back together. This is the basis of a number of approaches, and such approaches are referred to as the *directed shotgun*. For the directed shotgun to be viable method, one needs to know the location of some features of the DNA molecule to serve as landmarks. This is much like trying to look at a map of a large region: to determine the fine scale aspects of land that the map is intended to describe, one needs some feature that is detectable on the low resolution map (e.g. a mountain) and is present on all of the maps that are of finer resolution. We call these landmarks *markers*, a term we also use to refer to any known feature of a genome. The simplest markers indicate which chromosome is under consideration. There are 2 broad categories for methods aimed at finding the location of genes: *genetic mapping* and *physical mapping*. After more discussion of markers, we will first consider some of the approaches that have been used to construct physical maps, then we will consider genetic mapping.

3.2 Types of markers

At first, other than aspects of chromosomal organization and structure (such as the location of the centromere), genes were the only known markers. Genes with known locations are suboptimal markers for mapping genomes because they are often too spread out over the genome (for complex organisms, like mammals and flowering plants). In addition, it is often hard to distinguish between the different alleles based on the phenotype. For this reason, other markers have been identified. We will see that markers that take different values (or have multiple alleles) are crucial for genetic mapping. A marker is said to be polymorphic if this is the case.

3.2.1 Restriction fragment length polymorphisms (RFLPs)

When DNA is treated with a *restriction endonuclease* the DNA segment is cut in specific places depending on which restriction endonuclease is used and the nucleotide sequence of the DNA molecule. Different restriction endonucleases recognize different DNA subsequences and therefore give rise to different fragmentation patterns. If there is variation between individuals in the locations of the sites where the cutting takes place (i.e. polymorphisms in certain *restriction sites*), then DNA from different individuals will be broken up into different numbers of fragments. There are approximately 10^5 RFLPs in the human genome, and each has 2 alleles (either a cut takes place there or it doesn't). Since these markers only have 2 alleles they are often not useful for genetic mapping since everyone may have the same allele at some locus in a given study (i.e. they are not sufficiently polymorphic). In addition, the locations of these markers are too spread out to use for fine scale mapping. However, these markers are used for physical mapping in a manner that will be discussed below.

3.2.2 Simple sequence length polymorphisms (SSLPs)

Often there are repeated sequences of a certain unit (e.g. CACACA), and these can be polymorphic (moreover, there are often multi-allelic forms for these markers). The polymorphism in these markers lies in the number of units which are repeated. The biological source of polymorphism at these loci is thought to have arisen due to a phenomenon called replication slippage that can occur when single stranded DNA is copied during cell division. Due to this variation in DNA sequence length among people, one cannot just think of physical distance on a chromosome in terms of base pairs. Hence simple models for DNA in which the sequence is of some fixed length with different base pairs appearing in the various linearly arranged positions are inadequate for modeling some aspects of DNA.

There are 2 basic sorts of SSLPs: mini-satellites (or variable number of tandem repeats) and micro-satellites. The difference lies in the length of the repeat unit, with micro-satellites having the shorter unit (usually just 2, 3 or 4 nucleotide units that are repeated a number of times that varies across people). The alleles that someone has at a loci where a SSLP is located are determined via a southern blot. Micro-satellites are preferred because they are more evenly distributed throughout the genome and the genotyping is faster and more accurate. However, a very small fraction of these markers are located in the coding portion of some gene. This last feature can be an undesirable feature for a marker since we generally want to use markers that are unassociated with a phenotype (as this will violate the assumptions that lead to Hardy-Weinberg equilibrium, as we will discuss in Section 4.2).

3.2.3 Single nucleotide polymorphisms (SNPs)

There are many (over 1,000,000) locations on the genome in which there is a polymorphism at a single nucleotide. These locations are known as SNPs. While these have the same weakness as RLFPs, namely there are only 2 alleles, there are advantages to the use of these markers. First is the huge number of them, but more importantly, large scale genotyping can be automated (in contrast to the other 2 types of markers). It is now fairly routine to genotype subjects at hundreds of thousands of SNPs. Since a SNP can only take 2 values, the extent of polymorphism at a SNP is summarized by the frequency with which one observes the less frequent allele, which is called the minor allele frequency (the other allele is referred to as the major allele). Typically a SNP must have a minor allele frequency of at least 1% to qualify as a SNP. Automatic genotyping can be conducted using microarrays, a technology discussed in the final 4 chapters. Since the goal of using these markers is to determine the alleles for a subject at every location in the genome at which humans are known to differ, one can obtain the genome of a subject by determining the alleles someone has at this much smaller number of locations. These are rapidly becoming the most common type of marker used to conduct genetic mapping due to automatic genotyping, however many data sets exist that used SSLPs. Researchers frequently distinguish between a coding SNP, i.e. a SNP in a coding region of a gene from a non-coding SNP. While both types play a role in genetic mapping, coding SNPs are of great interest since variation in that SNP can lead to differences in the gene product (recall that some codons map to the same amino acid so not all variations in the nucleotide sequence lead to changes in the amino acid sequence).

3.3 Physical mapping of genomes

Many methods for physical mapping have been proposed, but there are 3 basic varieties:

1. restriction mapping,
2. fluorescent *in situ* hybridization (FISH),
3. sequence tagged site (STS) mapping.

Here we give the idea behind each, then look at the details of one method, namely, radiation hybrid mapping. This method falls under the category of STS mapping whose distinctive feature is that clones are assayed to determine if they have a certain set of loci.

3.3.1 Restriction mapping

This is a method for finding RFLPs. The simplest example is to first digest a DNA molecule with one restriction enzyme, then digest the molecule with another enzyme, then finally digest the molecule with both enzymes at the same time. After each digest, one separates the fragments using gel electrophoresis (a method that exploits the fact that smaller fragments will move further through a gel when exposed to some force to make the fragments move) to determine the length of the products after the digest takes place. After conducting each of these digests, one attempts to determine where the locations of the restriction sites for each enzyme are located. For example, if using restriction enzyme a yields fragments of length $a_1, \ldots, a_n$ and use of restriction fragment b yields fragments of length $b_1, \ldots, b_m$ then there must be $n - 1$ restriction sites for enzyme a and $m - 1$ restriction sites for enzyme b, but that is all that is known (note $\sum_i a_i = \sum_i b_i$ since we are using the same sequence for both digests). If we can find the relative locations of these $m + n - 2$ sites then we will have a physical map that indicates the locations of certain known features (here the restriction sites). Thus the goal is to use the information from the experiment where both restriction enzymes are used to find these relative locations. Suppose there are only 2 restriction sites (one for each enzyme), so that each digest using a single enzyme produces 2 fragments. Each of these indicate how close to one end of the sequence each restriction site is, yet does not indicate the relative location of these 2 sites. If we assume that a_1 is the distance from the end of the sequence we will call 0 to the location of the restriction site for enzyme a, then if $a_1 < b_1$ and $a_1 < b_2$ then the use of both enzymes will either result in the set of fragments of length $a_1, b_1 - a_1$ and b_2 or $a_1, b_2 - a_1$ and b_1. If we observe the first set of fragments then the restriction site for enzyme b must be a distance of b_1 from the end of the sequence that we have chosen to be zero. If we observe the second set of restriction fragments then the restriction site for enzyme b must be at location b_2 from the end of the sequence taken to be zero. While for this simple example, one can determine the distance between the restriction sites, for even slightly more complex examples it is impossible to determine the order from the restriction using both (or, more generally, all) of the restriction enzymes.

One can resolve any uncertainties by conducting a partial restriction, i.e., stop the restriction before it is complete. As simple examples illustrate, this analysis is very difficult if there are many restriction sites, so this method has limited usefulness (this method has been used to map virus genomes). Progress has been made by using restriction enzymes which cleave the DNA molecule at only a few sites (so called *rare cutters*). The computational problem of reassembling the sequence of the genome from DNA that has been digested by 2 restriction enzymes is known as the *double digest problem* (see Waterman, 1995 for a full treatment). This problem can be viewed as attempting to find the permutation of the indices for the restriction fragments that is most consistent with the data, and as such, can be thought of as a traveling salesman problem.

To understand the difficulty, consider the following simple example. Suppose that we use 2 restriction enzymes, call them A and B, and suppose that after applying A to a 7.9 kb (i.e. kilobases, or 1000 bases) sequence of DNA we obtain fragments of the length 3.7 and 4.2, while after applying B we get fragments of length 0.4, 1.8, 2.1 and 3.6. This indicates that the restriction site for enzyme A is 3.7 kb from one of the ends of the original fragment, but not much else. Now suppose we use both enzymes simultaneously and obtain fragments of length 0.4, 1.6, 1.8, 2.0 and 2.1. Now much more is known about the relative locations of these restriction sites. After some thought, we can discern 2 possible maps for the relative positions of the

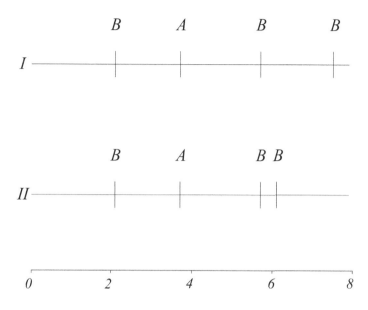

Figure 3.1 *The 2 possible maps of locations of restriction fragment sites that are consistent with the example discussed in the text.*

restriction sites based on these sets of fragmentation patterns, and these are shown in Figure 3.1. Now suppose we have many copies of this 7.9 kb fragment, and we run a partial restriction with enzyme A. This means that we stop the reaction before all of the fragments are cut by the restriction enzyme. If p_A is the probability that enzyme A cuts a fragment in this partial restriction (with p_B defined analogously), then we expect to observe a set of fragments where about $1 - p_A$ of the fragments are of length 7.9, about $p_a/2$ are of length 4.2 and $p_A/2$ are of length 3.7. This is because, for each fragment that gets cut, 2 fragments result and these 2 fragments will be found in exactly equal frequency. However the result is quite different if we use restriction enzyme B. To see this, we will suppose that each of the maps in Figure 3.1 is the true map, then examine the probability distribution of the length of the fragments. The result is shown in Table 3.1. We can see that each map has unique fragments that would not be observed if the other map was the true map.

Next note that if we let x represent the length of a random fragment that is sampled

Table 3.1 *Probabilities for observing lengths of fragments from a partial restriction.*

Map I		Map II	
probability	fragment lengths	probability	fragment lengths
p_B^3	0.4, 1.8, 2.1, 3.6	p_B^3	0.4, 1.8, 2.1, 3.6
$p_B^2(1-p_B)$	0.4, 1.8, 5.7	$p_B^2(1-p_B)$	0.4, 1.8, 5.7
$p_B^2(1-p_B)$	0.4, 2.1, 5.4	$p_B^2(1-p_B)$	1.8, 2.1, 4.0
$p_B^2(1-p_B)$	2.1, 2.2, 3.6	$p_B^2(1-p_B)$	2.1, 2.2, 3.6
$p_B(1-p_B)^2$	2.1, 5.8	$p_B(1-p_B)^2$	2.1, 5.8
$p_B(1-p_B)^2$	2.2, 5.7	$p_B(1-p_B)^2$	2.2, 5.7
$p_B(1-p_B)^2$	0.4, 7.5	$p_B(1-p_B)^2$	1.8, 6.1
$(1-p_B)^3$	7.9	$(1-p_B)^3$	7.9

after the partial restriction then we have

$$
\begin{aligned}
p(x) &= p(x|0\text{ breaks})p(0\text{ breaks}) + p(x|1\text{ break})p(1\text{ break}) \\
&\quad + p(x|2\text{ breaks})p(2\text{ breaks}) + p(x|3\text{ breaks})p(3\text{ breaks}) \\
&= p(x|0\text{ breaks})(1 - p_B)^3 + p(x|1\text{ break})p(1\text{ break}) \\
&\quad + p(x|2\text{ breaks})p(2\text{ breaks}) + p(x|3\text{ breaks})p_B^3.
\end{aligned}
$$

If we further suppose that the individual fragments lengths are measured with normally distributed error with variance σ^2 then we can get a more explicit expression for $p(x)$. First, under this condition we clearly have

$$
p(x|0\text{ breaks}) = \mathrm{N}(x|7.9, \sigma^2).
$$

With a little more work we find that

$$
\begin{aligned}
p(x|3\text{ breaks}) &= (1/4)\Big(\mathrm{N}(x|0.4, \sigma^2) + \mathrm{N}(x|1.8, \sigma^2) + \mathrm{N}(x|2.1, \sigma^2) \\
&\quad + \mathrm{N}(x|3.6, \sigma^2)\Big).
\end{aligned}
$$

This is because for each of the original fragments, we obtain 4 fragments and we measure the lengths of these fragments subject to normal errors. To find an expression for $p(x|1\text{ break})p(1\text{ break})$, note that there are 3 ways that 1 break can occur, and for each of these breaks 2 fragments will be produced. Each of these 2 fragments will occur with equal frequency and all breaks occur with the same probability, $p_B(1 - p_B)^2$. For each fragment that occurs from a single break, we will obtain a normal density, so we find that $p(x)$ is a linear combination of normal densities. The densities for both maps are plotted in Figure 3.2 for 4 choices of p_B. Note that when $p_B = 0.2$ almost all of the fragments are the length of the original fragment, whereas when $p_B = 0.8$ one can hardly distinguish between the 2 maps (the 4 peaks in that plot correspond to the lengths of fragments one would observe from the complete restriction). For intermediate cases, the 2 maps can be distinguished based on the observed fragment lengths.

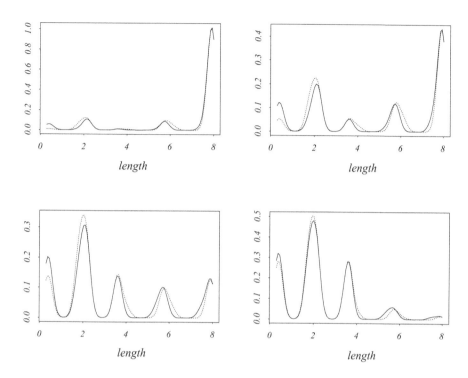

Figure 3.2 *An example of the densities of the lengths of fragments observed with error from a partial restriction for 2 different marker maps. Four different values of* p_B *shown: 0.2 (upper left), 0.4 (upper right), 0.6 (lower left), and 0.8 (lower right). Map 1 is represented by the solid line.*

3.3.2 *Fluorescent in situ hybridization (FISH) mapping*

This method is based on using a fluorescent dye to label a sequence of DNA, then creating conditions whereby the labeled sequence will bind to its complementary version in the genome under study. One then examines where the labeled probe binds to the chromosome. By using multiple probes with different dyes, one can determine the relative position of a number of DNA sequences. Originally, one would measure under the microscope how far the fluorescent label is from the end of the short arm of a chromosome (with the chromosome in a portion of the cell cycle known as metaphase). This procedure gave very poor resolution (although one could at least tell what chromosome a segment was on), but recent variations have improved resolution (by "stretching the chromosomes out," not using metaphase chromosomes or using purified DNA). Besides difficulties with resolution, FISH can only locate a few markers at once.

3.3.3 Sequence tagged site (STS) mapping

STS mapping is the most powerful physical mapping procedure. A *sequence tagged site* is a short segment of DNA (100 to 500 bp in length) that only occurs once in a chromosome or genome and is easily identified via a molecular probe. In this procedure, one first fragments a section of DNA into a set of overlapping fragments. Given a set of fragments, one then determines how close 2 STSs are by noting how often they are on the same fragment. SSLPs are often used as STSs as are random sequences of DNA and *expressed sequence tags* (ESTs). An EST is a segment of DNA which codes for a protein and occurs once in a genome. ESTs are typically 500-800 nucleotides in length. There are 2 basic methods for producing the collection of DNA fragments: radiation hybrids and clone libraries. Below we discuss radiation hybrids in detail. While construction of clone libraries differs from construction of radiation hybrid panels, the ideas for using either for STS mapping involve the same considerations.

3.4 Radiation hybrid mapping

3.4.1 Experimental technique

The basic idea of radiation hybrid mapping is to bind human chromosome fragments into a rodent (typically hamster) genome, then determine what markers are present in the same rodent cells. If 2 markers are present in the same rodent cell, they are most likely close on the human chromosome. Construction of a radiation hybrid panel starts with subjecting human cells to X-ray radiation. This causes the chromosomes to break apart (3000-8000 rads of radiation result in fragments of length 5-10 Mb). Then one stimulates the cells containing the fragmented human DNA to fuse with hamster cells. Since not all of the hamster cells take up the human DNA we get rid of these by growing the fused cells in a medium which selectively kills the cells without human DNA. In this way we can generate a whole genome radiation hybrid panel. Originally the method was used to map just a single chromosome at a time, but the details of that method imply the work required to map a whole genome is about the same as the work necessary to map a single chromosome, so the whole-genome method is what is used currently. Standard whole-genome radiation hybrid panels are now available and high resolution physical maps of the human genome have been produced (Olivier et al., 2001).

3.4.2 Data from a radiation hybrid panel

The data for producing an estimate of locus order or relative marker locations using radiation hybrid methods with m markers is an m-vector of zeros and ones for each hybrid cell. In these data vectors, a one indicates the human marker is present and a zero indicates otherwise. We will use y_i to denote the data vector for subject i and y

to denote the collection of these vectors. Note that when the goal is to estimate the order of a set of loci on the chromosome, the order of the markers as given in the vector y is assumed to be some arbitrary order. There is often missing data due to ambiguities in determining if the human marker is present or not.

3.4.3 Minimum number of obligate breaks

One criterion used to estimate the order of markers along a chromosome is to use the *minimum number of obligate breaks* method. To understand the method, suppose we observe a data vector for some hybrid cell like (0,0,1,1,0,0). There are a variety of breaking and retention patterns that could give rise to such data. For example, there could be a break between every marker (5 breaks) and sometimes the human marker is retained whereas other times the rodent DNA is retained. Another explanation is there are only two breaks: one between markers 2 and 3, and one between markers 4 and 5. If as low of doses of radiation are administered as is possible to break up the DNA, we would expect that the true number of breaks is as small as the data would allow. In the previous example, we would think the scenario involving 2 breaks is more likely than the scenario involving 5 breaks. In this case, we would say 2 is the minimal number of obligate breaks since at least this many are necessary to account for the observed retention pattern. Given a data vector for cell i with element $y_{i,j}$ for $i = 1, \ldots, n$ and $j = 1, \ldots, m$, we can compute the number of obligate breaks for this subject by counting how many times $y_{i,j} \neq y_{i,j+1}$. If 1_A is an indicator variable for the event A, then the number of obligate breaks for the ith hybrid is $B_i = \sum_j 1_{\{y_{i,j} \neq y_{i,j+1}\}}$.

Recall we have computed the number of obligate breaks using the ordering of the markers in some arbitrary order. To find the most likely order, we should compute this variable for each subject for every possible permutation of the markers. We denote permutations with σ. In the previous example, (0,0,1,1,0,0), if we use the permutation $\sigma = (1, 2, 5, 6, 3, 4)$ then we get the permuted data (0,0,0,0,1,1), and this vector has only one obligate break. The idea of the method is to find the permutation, σ, which makes the average number of obligate breaks $\frac{1}{n} \sum B_i(\sigma)$ as small as possible. For just a few markers, one can compute this average for every permutation and select the permutation giving the smallest average. When there are too many markers for enumeration to be feasible one can use approaches developed for the traveling salesman problem, as in Section 3.1.

Consistency of the order

Here we explain why the permutation of the marker order that minimizes the number of obligate breaks provides a consistent method for ordering markers. Recall that an estimate is said to be consistent for a parameter when the accumulation of more observations makes the estimate progressively closer to the true value of that parameter. To see that this estimate is consistent, first note that by the law of large numbers, our estimate converges to $EB_1(\sigma)$. Assume that the identity permutation $I = (1, 2, \ldots, m)$ is the true order of the markers (so our numerical ordering is the true order of the markers). This assumption is much like saying "suppose θ_0 is the true value of the unknown parameter" in cases where the parameter is a real number. Since the parameter space is finite (the set of permutations of a finite number of integers) if we just show

$$EB_1(\sigma) \geq EB_1(I),$$

then we will know our estimate is consistent. Since $B_1(\sigma) = \sum_j 1_{\{y_{i,\sigma(j)} \neq y_{i,\sigma(j+1)}\}}$, $EB_1(\sigma) = \sum_j P\{y_{1,\sigma(j)} \neq y_{1,\sigma(j+1)}\}$. To see why the identity permutation minimizes this expression consider the permutation $\sigma = (1, 2, 4, 3, 5)$. For this permutation

$$
\begin{aligned}
EB_1(\sigma) - EB_1(I) &= P(y_{1,2} \neq y_{1,4}) + P(y_{1,3} \neq y_{1,5}) \\
&\quad - P(y_{1,2} \neq y_{1,3}) - P(y_{1,4} \neq y_{1,5}).
\end{aligned}
$$

Suppose that every fragment is retained in the hybrid cells with probability ρ and there is a break between the ith and $(i+1)^{\text{th}}$ loci with probability θ_i. Then $P(y_{1,2} \neq y_{1,4}) = 2\rho(1-\rho)[\theta_2 + \theta_3] - 2\rho(1-\rho)\theta_2\theta_3$ whereas $P(y_{1,2} \neq y_{1,3}) = 2\rho(1-\rho)\theta_2$, hence $P(y_{1,2} \neq y_{1,4}) \geq P(y_{1,2} \neq y_{1,3})$ (since $\theta_2 \leq 1$). Clearly the same sort of reasoning will apply to the other difference for our example permutation. The basic idea expressed with this example is that if the true ordering follows our numerical ordering, then the expected value of $B_1(\sigma)$ for any permutation that deviates from the identity will entail the probability of an event like $y_{1,j} \neq y_{1,k}$ for $j - k > 1$. Such events have higher probability than $y_{1,j} \neq y_{1,j+1}$ since there is more than one opportunity for a break to occur. These higher probabilities make the expected value for any permutation other than the identity higher than it is for the identity permutation.

3.4.4 Maximum likelihood and Bayesian methods

The method presented above has two shortcomings. First, we do not obtain estimates of the distances between the the the loci. This is an important shortcoming from the perspective of building physical maps. Secondly, there is no way to assess the strength of evidence in favor of a given order. Likelihood based approaches can overcome both of these problems.

It suffices to consider how to compute the likelihood for a single hybrid (since all observations are assumed iid). For this, we will take advantage of the Markov property for Poisson processes. As mentioned above, we think of the breaks as occurring along the length of a chromosome according to a Poisson process. But this implies that breaks between the fixed locations given by the markers occur according to a Markov chain. This implies that we can write the likelihood for a single observation as

$$L(y_i|\theta,\rho) = L(y_{i,1}) \prod_{j=2}^{m} L(y_{i,j}|y_{i,j-1},\ldots,y_{i,1})$$

$$= L(y_{i,1}) \prod_{j=2}^{m} L(y_{i,j}|y_{i,j-1})$$

Now $L(y_{i,1})$ is just the likelihood for a binomial with likelihood of success given by the retention probability ρ. For now, suppose the order of the loci is known and is the same as the numerical ordering we are using. To find $L(y_{i,j+1}|y_{i,j})$ we note $y_{i,j+1}$ is also Bernoulli, and if, for example, $y_{i,j} = 0$ then $y_{i,j+1}$ is a success with probability $\rho\theta_j$ (i.e. a break in a fragment at that location and retention of the human fragment). Similarly, if $y_{i,j} = 1$ then $y_{i,j+1}$ is a success with probability $\rho\theta_j + 1 - \theta_j$ (i.e. a break in a fragment at that location and retention of the human fragment or no break). We then find that we can write the likelihood for a single cell as

$$L(y|\rho,\theta_1,\ldots,\theta_{j-1}) = \prod_{i=1}^{n} \rho^{y_{i,1}}(1-\rho)^{1-y_{i,1}} \prod_{j=1}^{m-1} \left(\rho\theta_j + y_{i,j}(1-\theta_j)\right)^{y_{i,j+1}}$$

$$\times \left(1 - \rho\theta_j - y_{i,j}(1-\theta_j)\right)^{1-y_{i,j+1}}.$$

One can then maximize the likelihood with respect to the parameters θ_j and ρ. To do so, note that if one differentiates the log-likelihood with respect to θ_j the result is

$$\frac{\partial \ell}{\partial \theta_j} = \sum_i \left[y_{i,j+1} \frac{\rho - y_{i,j}}{\rho\theta_j + y_{i,j}(1-\theta_j)} + (1 - y_{i,j+1}) \frac{y_{i,j} - \rho}{1 - \rho\theta_j - y_{i,j}(1-\theta_j)} \right].$$

If we then let n_{jk} represent the number of times that $y_{i,l} = j$ and $y_{i,l+1} = k$ we find that

$$\frac{\partial \ell}{\partial \theta_j} = n_{00}\rho/(\rho\theta_j - 1) + (n_{01} + n_{10})/\theta_j + n_{11}(\rho - 1)/(\rho\theta_j + 1 - \theta_j).$$

If we then set this partial derivative to zero and solve for θ_j, we find that we have the following quadratic equation in θ_j that only involves ρ, and n_{jk} for $j = 0,1$ and $k = 0,1$:

$$\theta_j^2\left(n\rho(1-\rho)\right) + \theta_j\left(\rho n_{00} + n_{10} + n_{01} + (1-\rho)n_{11}\right) - (n_{01} + n_{10}) = 0.$$

Hence one can obtain an explicit expression for the MLE of θ_j, denoted $\hat{\theta}_j$, given by the expression

$$\hat{\theta}_j = \frac{\rho n_{00} + n_{10} + n_{01} + (1-\rho)n_{11}}{2n\rho(1-\rho)}$$

$$- \frac{1}{2}\sqrt{\left[\frac{\rho n_{00} + n_{10} + n_{01} + (1-\rho)n_{11}}{n\rho(1-\rho)}\right]^2 - \frac{4(n_{01} + n_{10})}{n\rho(1-\rho)}}.$$

The lower root of the quadratic equation is usually less than 1, hence we have used this root here. However, there is no guarantee that this root will be in the interval

$[0, 1]$. This can happen if $n_{01} + n_{10}$ is large relative to n. For example, if $n_{01} + n_{10} = n$ and we set $\rho = \frac{1}{2}$ we find that $\hat{\theta}_j = 2$ in the previous expression and the likelihood is maximized over the interval $[0, 1]$ at the point 1. In this case the usual theory for MLEs fails to hold, however the Bayesian analysis remains unchanged (the posterior mode is at 1). In any event, note that $\hat{\theta}_j$ depends just on ρ and not θ_k for any $k \neq j$. Thus one can find the MLEs here by substituting $\hat{\theta}_j(\rho)$ into the log-likelihood and conducting a univariate maximization to obtain $\hat{\rho}$ (which can be done by simply plotting the log-likelihood as a function of ρ). Once $\hat{\rho}$ is determined, we can substitute this value into $\hat{\theta}_j(\rho)$ to obtain the MLEs of the set of θ_j.

We can also use likelihood based methods to find the most likely order or to determine if there are several orderings that are nearly equally likely. First, we can in principal find the MLE of the order by considering every possible order, find the MLEs of all parameters, then substitute these estimates back into the likelihood. The MLE of the order is then given by the order that maximizes the resulting expression. While we can look at ratios of log-likelihoods to compare the strength of evidence in favor of certain orderings, the Bayesian approach allows one to compute ratios between posterior probabilities in favor of certain orderings, and these are easily interpreted. One can more simply compute probabilities for many orders and retain those which are sufficiently high. In practice most of the posterior probability is confined to a short list of orders with appreciable probabilities (see Lange and Boehnke, 1992). For this reason, some authors (such as Lange, 1997) advocate the use of the minimum number of obligate breaks to determine the order of the loci, then use likelihood based methods to determine the relative distance between the loci.

A characteristic of these data sets is that there are frequently missing values that arise due to ambiguity in the measured data. To account for this missing data, we use the EM algorithm treating the missing data as missing. We can use the previous expression for the likelihood to obtain an expression for the complete data log-likelihood. The E step then consists of taking the expectation of this expression conditional on current values for the retention probabilities and ρ. The M step is then just the same as finding the MLEs when there is no missing data. More on these methods can be found in Lange and Boehnke (1992), Lunetta et al., (1996) and Boehnke, Lange and Cox (1991).

3.5 Exercises

1. Consider a genome of length G basepairs. Suppose that one has a set of N fragments of DNA of identical length L. Further suppose that the start locations of these fragments are iid. Show that the proportion of the genome that is covered by this set of fragments is approximately

$$1 - e^{-N\frac{L}{G}}$$

   when L and N are small relative to G.

2. Suppose we treat a 5.7 kb length sequence on which we would like to know the

locations of the restriction sites for 2 restriction endonucleases, A and B. When we treat the sample with A, we get products of length 2.6, 1.7 and 1.4. When we treat the sample with B, we get 2.7, 2.0, and 1.0. When we use both, we get 2.0, 1.7, 1.0, 0.6, and 0.4.

(a) List every set of loci for the restriction sites which are consistent with the fragmentation patterns.

(b) To clear up ambiguities in the order we should run a partial restriction. Suppose the probability of breakage, p, is the same for all restriction sites. If there length of the measured fragments is measured with error so that we can approximate the lengths with a normal random variable with mean given by the true length and standard deviation of 0.2, what is the distribution of the fragment lengths for each of the orderings you listed in part A? Produce histograms of these distributions.

(c) Use your answer from part (b) to explain which partial restriction you should run, A or B.

3. Use the Markov property and the definition of conditional probability to show we can write the likelihood for radiation hybrid data as

$$L(y|\rho, \theta_1, \ldots, \theta_{j-1}) = \prod_{i=1}^{n} L(y_{i,1}) \prod_{j=1}^{m-1} L(y_{i,j+1}|y_{i,j}).$$

4. Derive an expression for the covariance matrix of the MLEs for a radiation hybrid panel data set when there are 3 markers.

5. First show that if y_{ij} is Markov in the parameter j then

$$p(y_{ij}|y_{i,j-1}, y_{i,j+1}) \propto p(y_{i,j+1}|y_{ij})p(y_{ij}|y_{i,j-1}).$$

Then use this to show how to implement the EM algorithm for radiation hybrid panel data when information for one marker is missing, but we observe the outcome for the flanking markers.

6. Suppose that we treat the order of the markers, σ, as a parameter. Describe how one could find the MLE of σ. If we suppose that all orders are equally likely, describe how one could sample from the posterior distribution of σ (assume that all probabilities that appear in the model are uniformly distributed on (0,1)).

7. In the context of the model proposed for radiation hybrid panel data, suppose that ρ and θ_j have uniform priors. Also, treat the permutation σ as a parameter. Show that the marginal posterior of ρ and σ satisfies

$$p(\rho, \sigma|y) \propto \rho^{\sum_i y_{i,\sigma(1)}} (1 - \rho)^{n - \sum_i y_{i,\sigma(1)}} \prod_{ij} \eta_{i,\sigma(j)}^{y_{i,\sigma(j+1)}} (1 - \eta_{i,\sigma(j)})^{1 - y_{i,\sigma(j+1)}}$$

where

$$\eta_{i,j} = \frac{1}{2}(\rho + y_{i,j}).$$

8. Suppose that when $y_{i,j} = 1$, $y_{i,j+1} = 1$ only if there is not a break that occurs between the 2 markers (instead of if there is not a break or if there is a break

and the human fragment is retained as we had assumed in Section 3.4.4). Find an expression for the MLE of θ_j as it depends on ρ. Also find an approximation to the variance of this estimate given some $\hat{\rho}$.

CHAPTER 4

Basic Linkage Analysis

4.1 Production of gametes and data for genetic mapping

In order to understand the basis for genetic mapping of disease, we need to consider the events that take place during *meiosis*, the formation of sex cells, also referred to as *gametes*. Sex cells (i.e. egg and sperm cells), unlike other cells in the body (referred to as *somatic cells*) only carry one copy of each chromosome, either the mother's or father's. Meiosis starts when a cell carrying 2 copies of each gene duplicates all of the chromosomes in the genome. Once the chromosomes have been duplicated, the cell divides into 2 cells with 2 copies of each chromosome. Then these 2 cells both divide one more time, resulting in 4 cells with 1 copy of each chromosome. Each of these 4 cells has a mix of chromosomes from the 2 parental sources so that each chromosome is equally likely to be found in the resulting 4 cells. When 2 sex cells fuse, the result then has a pair of chromosomes.

During meiosis, after all the DNA is duplicated, the chromosomes with the same number but from the different parental sources pair up and at certain random locations on the chromosomes contact one another. When the chromosomes touch (which is known as a *crossover*), they exchange segments of DNA, thereby creating a pair of chromosomes which are distinct from either of the progenitor chromosomes. This process is called *recombination*, and was first observed in the early twentieth century. The process of recombination is thought to be driven by chromosomal breakage: a break in one chromosome is repaired by using the other chromosome in the pair as a template, and this results in copying DNA from one chromosome to the other in a pair. Actually, since multiple crossovers are possible, not all crossovers will lead to 2 alleles on opposite sides of one crossover being on distinct chromosomes, hence recombination is reserved for the cases where 2 alleles end up on different chromosomes than where they started. If 2 loci are close on the chromosome, it is unlikely they will be separated by a crossover since these occur randomly along the genome (although they do not occur uniformly over the genome). The *recombination rate* between 2 loci is the probability of the 2 loci having alleles from different parental sources. If 2 loci are so close that it is highly unlikely that they will be separated by a crossover, then the recombination rate would be near zero. If the 2 loci are on distinct chromosomes, then they will have alleles from the same parental source if the gamete happened to get the 2 chromosomes from the same parental source, and this would occur with probability 0.5 (since gametes have a single set of chromosomes where the parental source of each chromosome is equally likely). The recombination rate is also thought of as the probability of a crossover occurring between 2 loci (which is difficult to interpret for loci on distinct chromosomes). This is measured in units called *Morgans* (after the scientist in whose lab crossing over was first observed). For genes on distinct chromosomes this is 0.5, but for genes on the same chromosome this is some quantity less than or equal to 0.5.

The data for genetic mapping in humans are pedigrees, genotypes and phenotypes. Since the underlying random process that gives rise to information regarding gene location is only observed through matings (i.e. meioses that result in offspring) we need data from related individuals to estimate the recombination rate. For experimental animals, we can design experiments to construct desired pedigrees, but this will not work for humans. There are a number of conventions used in writing down a pedigree (see Figure 4.1 for an example). Each individual is represented as either a circle (for females) or a square (for males). Horizontal lines between individuals indicate a mating, and vertical lines drawn down from these horizontal lines indicate offspring. Typically the diseased individuals are indicated by a darkened simple and a deceased individual has a line drawn through his or her symbol. Monozygotic twins are represented as the lower vertices of a triangle. Sometimes numbers are used next to individuals to represent genotypes, and if an individual's genotype is not measured but can be unambiguously identified, it is put in parentheses. Next, as we'll see in Section 4.3, a parent who is homozygous for the marker can't tell you anything about recombinations, hence the mates of homozygous affected parents are sometimes excluded from the pedigree. Finally, individuals whose parents are not included in the pedigree are referred to as *founders*. There are other symbols used to indicate things like spontaneous abortions and if an individual is the proband (i.e. the first affected family member to participate in the study).

An important distinction from the perspective of computing the recombination rate is between *simple* and *complex* pedigrees. A simple pedigree has no closed loops, whereas a complex pedigree does. Complex pedigrees are quite common in animal populations (especially domesticated, breeding animals like horses) but are less common in human families (although there are cultures in which matings between first cousins are common, and this can give rise to complex pedigrees).

4.2 Some ideas from population genetics

Population genetics is a field concerned with population level phenomena relating to genetics. For example, studying the frequencies of certain alleles in a species over time and relating this to the ability of the species to survive is a topic of importance. An important notion from population genetics that we will need is the idea of Hardy-Weinberg equilibrium. Put succinctly, genotype frequencies in the population depend only on gene frequencies. To make this statement rigorous requires a number of assumptions. In particular, we need to assume:

1. infinite population size,

2. discrete generations,

3. random mating with respect to the allele (also known as *no selection*),

4. no migration,

5. no mutation,

6. no association between the genotypes and sex.

Clearly some of these assumptions are likely to not hold for many traits of interest (the first 2 are more technical assumptions required for the proof). On the other hand, they frequently hold for genetic markers since most markers are not even in genes. Under these assumptions, a population will reach Hardy-Weinberg equilibrium in one generation. To see this, suppose some gene has 2 alleles, A_1 and A_2, and these occur in the population of gametes with probabilities p_1 and p_2. The list of assumptions given above implies independent combination of the gametes, hence the genotypes occur with the following probabilities $P(A_1, A_1) = p_1^2$, $P(A_1, A_2) = 2p_1p_2$, and $P(A_2, A_2) = p_2^2$. Now consider the population of gametes in the next generation. We find $P(A_1) = p_1^2 + p_1p_2 = p_1$ and $P(A_2) = p_2^2 + p_1p_2 = p_2$. This argument easily generalizes to the case of more than 2 alleles. Just from the finiteness of human populations we expect gene frequencies to change randomly over time, a phenomenon refereed to as *genetic drift*. By use of a χ^2 goodness of fit test we can test the null hypothesis of Hardy-Weinberg equilibrium by comparing the observed allele frequencies with the observed genotype frequencies.

There is a set of techniques, referred to as *segregation analysis*, which aims to determine if a disorder is a recessive trait or a dominant trait. We will see that knowledge of the mode of inheritance is important when computing genetic distances, so this question is of considerable interest. If a trait is recessive, only 25% of offspring from heterozygous parents would be affected, while if the trait is dominant 50% of the offspring would be affected. For X-linked traits there are also distinctive patterns of segregation. Based on the expected ratios of affecteds amongst offspring, with experimental animals one can devise experiments to determine which mode of inheritance provides a better model for the data, but for human populations, the situation is more complex. Typically one must model the mechanism by which an affected individual is *ascertained* (i.e. made a member of the study population). Typically families are ascertained on the basis of a single affected individual, the proband. Since the results will depend on the ascertainment model, many researchers claim relying on the results of a segregation analysis is not helpful. Instead, one should just fit the recessive and the dominant model and let the data select the appropriate model (see, for example, Ott, 1999). Alternatively one can use the methods presented in Chapter 6.

4.3 The idea of linkage analysis

In radiation hybrid mapping, we were able to determine that 2 markers are close by observing how frequently they are on the same fragment of DNA after subjecting a genome to X-ray radiation. Linkage analysis is based on the same idea, except that recombination during meiosis takes the place of X-ray radiation, and we observe the phenotype of each individual. Since we observe the phenotype, we can try to deduce not only the location of a gene, but also its function. Assume each phenotype arises due to a specific genotype for some gene. Under this assumption, by examining what marker alleles are shared by individuals with the same phenotype, we can try to locate the gene responsible for the phenotype (since recombination will usually not separate a disease allele from marker alleles nearby it on a chromosome). One problem with this approach is that affected subjects in a family will usually share the same alleles over a segment of DNA that contains many genes. We will return to this idea in Chapter 6 when we discuss fine mapping.

As a simple example, suppose we are studying a dominant disorder and have some data on a family. In particular, suppose one of the parents is affected, as is an offspring (the other parent is unaffected). If we assume that all affecteds are heterozygous at the disease gene and all unaffecteds are homozygous at the disease gene (with the healthy allele), then we know the genotype at the disease locus for all subjects. Suppose we genotype all subjects at some marker locus, and both parents are homozygous at this marker but with different alleles. Suppose that the affected parent has 2 copies of the allele we label 1 and the unaffected parent has 2 copies of the allele we label 2. In this case, the offspring will necessarily be heterozygous at the marker locus (assuming no mutation) and we know which alleles the offspring received from each parent (i.e. we know the offspring's haplotype). If the offspring then mates with someone who is unaffected and has different alleles at the marker locus (say, 2 copies of allele 3), we can determine if a recombination has occurred between the marker locus and the disease locus by examining the resulting offspring's marker alleles and affectation status. For example, if the affected offspring in this third generation has marker allele 1, then we know a recombination did not occur between the marker locus and the disease locus since a recombination would create a sex cell that has the disease allele and marker allele 2. Conversely, if an unaffected offspring has allele 1 then a recombination must have occurred during formation of the sex cell that led to offspring in the third generation. If we saw many families where no recombination occurred between the disease locus and some marker locus we would then think that the disease locus must be right next to that marker locus. In this sense, we have determined the location of a gene and its function simultaneously. Unfortunately, one is frequently unable to determine if a recombination has occurred between 2 loci, as illustrated in the next example.

As a more complex example, consider the pedigree in Figure 4.1 with marker alleles at one locus as given in Table 4.1. This represents a 3 generation family, and the figure was generated with the program MADELLAINE 2 (described in Trager et al., 2007), a flexible, free to use program for drawing pedigrees. Here all subjects have been sampled and subject U00106 is the proband (as indicated by the arrow). Suppose that the disease acts in a dominant fashion. If we use D to denote the disease allele and d to denote the non-disease allele, then the genotype for the mother is either (D, d) or (D, D). We also usually assume that the disease allele is sufficiently rare

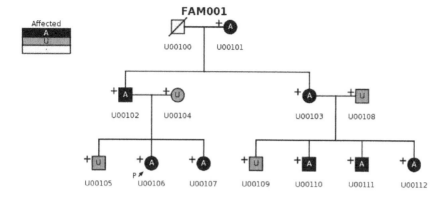

Figure 4.1 *A pedigree for a family with may affecteds. We assume the disorder is dominant.*

Table 4.1 *Marker data for FAM001.*

subject	marker alleles
U00100	2,2
U00101	1,2
U00102	1,2
U00103	1,2
U00104	1,2
U00105	1,2
U00106	1,1
U00107	1,2
U00108	1,2
U00109	1,2
U00110	2,2
U00111	1,1
U00112	1,1

so that we can ignore the (D, D) genotype (an assumption frequently referred to as the rare dominant assumption). If we then assume all healthy individuals have the non-disease genotype (d, d), we know the genotypes at the disease locus for all subjects. Next note that even though we don't know which 2 alleles are on the same chromosome (i.e. the phase) we can deduce this for our disease gene and the marker gene for almost all subjects. First note that all unaffected subjects are homozygous at the disease gene, hence we know the phase for all these subjects. For example, if we denote phased genotypes as $(a_1, b_1 | a_2, b_2)$ where a_1 and a_2 are the alleles for loci 1,

b_1 and b_2 are the alleles for loci 2, and a_1 and b_1 are on the same chromosome, then subject U00100 has a phased genotype $(d, 2 | d, 2)$, and subject U00104 has phased genotype $(d, 1 | d, 2)$.

In fact, it is only the doubly-heterozygous subjects (i.e. U00101, U00102, U00103, and U00107) whose phase is not obvious. To determine phase for these subjects, consider subject U00101, and suppose her phased genotype were $(D, 1 | d, 2)$. Then, if no recombination occurs during the formation of a sex cell, the sex cells would have genotypes $(D, 1)$ and $(d, 2)$ (recall that sex cells only have one chromosome). The gametes of subject U00100 will all have genotype $(d, 2)$, so the phased genotype of the offspring will just be determined by U00101. This genotype will be either $(D, 1 | d, 2)$ or $(d, 2 | d, 2)$, and each is equally likely. If a recombination did occur, then the gamete genotypes would be $(D, 2)$ and $(d, 1)$ which would result in the phased genotypes $(D, 2 | d, 2)$ and $(d, 1 | d, 2)$. On the other hand, if we suppose the mother has phased genotype $(D, 2 | d, 1)$ then we find that if there is no recombination the phased genotypes of the offspring are equally likely to be $(D, 2 | d, 2)$ and $(d, 1 | d, 2)$, while if there is a recombination the phased genotypes are equally likely to be $(D, 1 | d, 2)$ and $(d, 2 | d, 2)$. Notice that for this example, the phased genotype of subjects U00102 and U00103 are known since both the disease allele and allele 1 must have come from their mother (since the father only has 2 copies of allele 2). This implies that the mother's phased genotype was either $(D, 1 | d, 2)$ and no recombinations occurred or her genotype was $(D, 2 | d, 1)$ and 2 recombinations occurred. Since the phase of subjects U00102 and U00103 is known we can directly count the number of recombinations among their offspring. Of the 5 meioses that led to an affected offspring, 3 involved no recombination, one subject (U00110) was conceived by an egg that had experienced a cross over during its development and one subject (U00107) is consistent with either a cross over or no cross over (so this subject is uninformative). Next, consider the unaffecteds, specifically subject U00105 (the same applies to U00109). Note that the phase of both parents is known. This subject's genotype can arise from either the father having a recombination or not having a recombination, thus these 2 subjects are also uninformative.

Thus of the 9 matings in this pedigree, 3 are uninformative, 3 involved no recombination, 1 involved a recombination, and the final 2 were either both recombinations or both were non-recombinations (depending on the unknown phase of U00101). One could argue that the 2 unsure outcomes should just count as nonrecombinants given the low recombination rate for the rest of the meioses, but a more rigorous approach is to average over the probability of each of the phases. If we use $P(D, 1 | d, 2)$ to represent the probability of this phased genotype in the population from which the founders come, then if we assume that in the founder population there is no association between any allele and the disease allele (a condition called linkage equilibrium that will be discussed further when association mapping is discussed in Chapter 6), then $P(D, 1 | d, 2) = P(D, d)P(1, 2)$ where $P(a, b)$ is the probability of observing the pair of alleles a and b in the population. But this implies both genotypes are equally likely, so that the number of recombinations obtained after averaging over the unknown phase is $0 \times 0.5 + 2 \times 0.5 = 1$. So we have a total of 4 matings

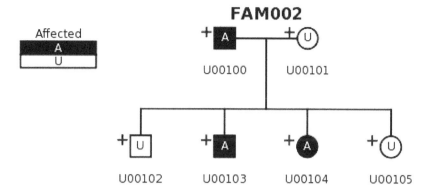

Figure 4.2 *A non-informative mating.*

Table 4.2 *Marker data for FAM002.*

subject	marker alleles
U00100	1,1
U00101	1,2
U00102	1,2
U00103	1,2
U00104	1,1
U00105	1,2

involving no recombination and 2 involving a recombination and an estimated re-combination rate of $2/6 = 0.33$. Finally, note that if we assume the conditions for Hardy-Weinberg equilibrium for the marker (which is reasonable if the marker is in a noncoding region of the genome) then $P(1,2) = P(1)P(2)$, and if we assume these conditions for the disease (which may not be reasonable if the disease is related to mating choice) we also have $P(D,d) = P(D)P(d)$. Since $P(1), P(2), P(D)$ and $P(d)$ are unknown, these must be treated as parameters in a full likelihood based approach.

As another example, consider the pedigree in Figure 4.2 with marker alleles as given in Table 4.2. If one considers the population of gametes that are potentially con-tributed by each parent it is clear that no mating here would ever give rise to offspring that are informative about whether a recombination occurred during meiosiss. Con-sider the population of gametes in the father. Suppose the disease acts in a dominant

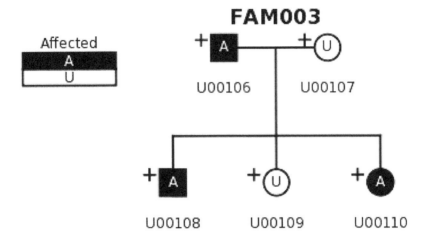

Figure 4.3 *A family affected by a trait that is assumed to be recessive.*

Table 4.3 *Marker data for FAM003.*

subject	marker alleles
U00106	1,3
U00107	1,2
U00108	1,3
U00109	2,3
U00110	1,1

fashion and is rare. Then his genotype at the disease locus must be (D, d). Then, if his phased genotype is $(D, 1|d, 1)$, if there is no recombination, the population of gametes will have haplotypes of either $(D, 1)$ or $(d, 1)$. If there is a recombination, then the population of gametes will also have haplotypes $(D, 1)$ or $(d, 1)$. Thus, since a recombination won't change the population of gametes, this mating can't be informative about whether or not a recombination has occurred.

Finally, Figure 4.3 shows a pedigree for a disease that acts in a recessive fashion, and Table 4.3 displays the alleles at the marker locus. Since the disease acts in a recessive fashion, all affecteds must have genotype (D, D) at the disease locus. This implies that the mother must have genotype (D, d) since the offspring must have received a disease allele from her (she can't have genotype (D, D) since she is unaffected). Also note that subject U00109 must have genotype (D, d) at the disease locus because the affected father always transmits a disease allele. In fact, for this example it is easy to determine the phased genotypes. First all affecteds are homozygous at the disease locus, hence determining phase is automatic. Subject U00109 must have received the marker allele 3 from the father, hence her phased genotype is just $(D, 3|d, 2)$. This implies that if no recombination took place during the creation of the egg that ultimately resulted in subject U00109, the mother has the phased genotype $(D, 1|d, 2)$. If there was a recombination then the phased genotype is $(D, 2|d, 1)$.

4.4 Quality of genetic markers

As noted above, if a parent is homozygous for a marker, we cannot use this parent to help localize a disease gene regardless of the genetic mode of transmission (i.e. dominant or recessive). For this reason, when we construct genetic maps, we want to use markers which are highly polymorphic. This is the motivation for using microsatellites. There have been a number of attempts to quantify the polymorphism of a marker. In practice, a marker may be highly polymorphic but may not be useful for other reasons, for example, the genotyping may be difficult due to problems with the genotyping technology for that marker.

4.4.1 Heterozygosity

The *heterozygosity*, H, is defined as the probability that an individual is heterozygous at a loci. If p_i represents the probability that an individual has allele i at the loci, then

$$
\begin{aligned}
H &= P(\text{heterozygous}) \\
&= 1 - P(\text{homozygous}) \\
&= 1 - P(\cup_i \text{homozygous with allele } i) \\
&= 1 - \sum_i P(\text{homozygous with allele } i) \\
&= 1 - \sum_i p_i^2.
\end{aligned}
$$

Inbreeding doesn't alter the p_is (recall we didn't assume no inbreeding in establishing Hardy-Weinberg equilibrium), but it does reduce the heterozygosity by a factor (commonly denoted $(1 - F)$). In human populations, inbreeding is uncommon so this can factor can be ignored. In exercise 1 we will see that heterozygosity is highest when all alleles occur with the same frequency. Under this condition we can find how many alleles are necessary to reach a certain level for the heterozygosity.

If we estimate H with the sample proportions, $\hat{p}_i$, then we will obtain a biased estimate of H (denoted $\hat{H}$), although this bias will be negligible in the limit (since plugging in the sample proportions provides the MLE of H and these are asymptotically unbiased). It is not hard to show that an unbiased estimate can be obtained by using $\frac{n}{n-1}\hat{H}$.

4.4.2 Polymorphism information content

Another measure of marker polymorphism is the *polymorphism information content* (PIC) (Botstein et al., 1980). It was devised as a measure of polymorphism for the case in which we are studying a rare dominant disease. The PIC is defined as the probability that the marker genotype of a given offspring will allow deduction of which of the 2 marker alleles of the affected parent it had received. The motivation for this measure is we want to measure the quality of a marker with regard to how helpful it is in deducing if a recombination occurred (since we want to determine which marker allele came from the affected parent). In terms of the allele probabilities, this can be expressed

$$\text{PIC} = 1 - \sum_i p_i^2 - 2\sum_i \sum_{j>i} (p_i p_j)^2.$$

For small p_i this will be approximately equal to H, but in general $\text{PIC} < H$.

To see why PIC is a better indicator of the informativeness of a mating, consider a mating that results in a single offspring. Suppose that one parent is affected with a dominant disorder as well as the offspring (so both affected subjects are heterozygous at the disease locus while the unaffected parent is homozygous with 2 copies of the healthy allele). Suppose a single marker with 2 alleles is typed in all 3 subjects and all 3 are heterozygous at the marker locus. In this setting, one can not unambiguously determine which marker allele came from the affected parent and so one can not determine if a recombination has occurred between the disease locus and the marker locus. Note that this ambiguity occurs even though all subjects are heterozygous at the marker locus and the affected parent is doubly heterozygous. PIC accounts for this possibility when quantifying the informativeness of a marker while H does not. This example also illustrates that while one parent being doubly heterozygous is a necessary condition for a mating to be informative about whether a recombination has occurred, this condition is not sufficient.

4.5 Two point parametric linkage analysis

We will now more formally consider the simplest case of linkage analysis, namely, there is a single disease gene, the disease is either recessive or dominant, and we have a single marker. This situation is referred to as *two point parametric linkage analysis*. If we consider multiple markers, then we refer to *multipoint linkage analysis*. Non-parametric linkage analysis is used to refer to methods which don't require specification of the mode of transmission (i.e. recessive or dominant).

4.5.1 LOD scores

Although geneticists use the techniques of maximum likelihood and the associated ideas regarding hypothesis testing (e.g. likelihood ratio tests), the specialized terminology used by geneticists makes this less than evident. A hypothesis of interest to geneticists is the null hypothesis of no linkage, which is expressed as

$$H_0 : \theta = 0.5.$$

If we reject this hypothesis the loci are said to be *linked*. A natural way to test this is to use a maximum likelihood ratio test. If $L(y|\theta)$ denotes the likelihood for some family data y, the test statistic is then $L(y|\theta)/L(y|1/2)$ maximized over θ. Geneticists don't use this test statistic, instead they use the *LOD score*, which is defined to be the base 10 logarithm of the maximized likelihood ratio. We often express the log-likelihood ratio as a function of θ:

$$Z(\theta) = \log_{10} \frac{L(y|\theta)}{L(y|1/2)}.$$

We then maximize this expression with respect to θ to obtain the MLE of θ. If we then substitute in the maximized value of θ we obtain what is known as a LOD score. Usually this maximization cannot be done directly (since phase is unknown), hence the methods considered in Section 4.7 must be used. The criterion for linkage is a LOD score greater than 3.0. If the χ^2 approximation holds for 2 times the natural logarithm of the log-likelihood ratio, then a LOD score of 3.0 implies we use an α-level of 0.0002 for our hypothesis test. While this may seem too strict, it is sensible given that many markers are frequently used in what are called genome scans and this raises the problem of multiple hypothesis testing.

4.5.2 A Bayesian approach to linkage analysis

As an alternative to just using LOD scores, we can use a Bayesian approach. Following Smith (1959), we suppose that prior to observing the data, there is a 1/22 chance that the disease gene is linked to the marker. This is a sensible assumption for markers and genes on autosomal chromosomes (since there are 22 of them). A more sophisticated approach could be developed that would make the prior probability of linkage depend on the number of genes on a chromosome. Our prior distribution can then be written

$$p(\theta) = \frac{1}{11} 1_{\{\theta < \frac{1}{2}\}} + \frac{21}{22} \delta_{\{\theta = \frac{1}{2}\}},$$

where $\delta_{\{\theta = \frac{1}{2}\}}$ is the Dirac delta function centered at $\theta = \frac{1}{2}$. If we use y to represent the family data, then we are interested in computing $p(\theta|y)$ since we can then easily find $P(\theta < 0.5|y)$. If we can compute this probability then we can at least determine if the marker and the gene are on the same chromosome (i.e. linked). Using Bayes theorem

$$
\begin{aligned}
p(\theta|y) &= \frac{p(y|\theta)p(\theta)}{\frac{1}{11}\int_{0\leq\theta<\frac{1}{2}} p(y|\theta)\,d\theta + \frac{21}{22}p(y|\theta=\frac{1}{2})} \\
&= 22\frac{p(y|\theta)p(\theta)/p(y|\theta=\frac{1}{2})}{2\int_{0\leq\theta<\frac{1}{2}} p(y|\theta)/p(y|\theta=\frac{1}{2})\,d\theta + 21}.
\end{aligned}
$$

If we then integrate this expression over the region $\theta < \frac{1}{2}$ we find

$$
P\left(\theta < \frac{1}{2}\ \middle|\ y\right) = \frac{2\Lambda}{2\Lambda + 21},
$$

where

$$
\Lambda = \int_{0\leq\theta<\frac{1}{2}} p(y|\theta)/p\left(y\ \middle|\ \theta=\frac{1}{2}\right)\,d\theta.
$$

If we assume phase is known, then the family data is just a count of recombinants out of a certain number of meioses, hence the binomial model is appropriate for the likelihood. If there are y recombinations out of n meioses

$$
p(y|\theta) = \binom{n}{y}\theta^y(1-\theta)^{n-y}.
$$

This implies

$$
\Lambda = 2^n \int_0^{\frac{1}{2}} \theta^y(1-\theta)^{n-y}\,d\theta.
$$

The integral here is known as the incomplete Beta function and there are a number of ways to compute such expressions. The most direct (since y and $n - y$ are integers) is to just use the binomial theorem, then integrate the resulting polynomial in θ. If phase is unknown, as is usually the case, then computation of the posterior probability of linkage must average over the unknown phase as in the second example in Section 4.3. In general one must use sophisticated algorithms to compute the likelihood as discussed in Section 4.7.

4.6 Multipoint parametric linkage analysis

We saw that the objective of 2 point linkage analysis was to localize a disease gene. By computing the genetic distance between many markers and genes or pairs of markers, we could localize a disease gene or construct a genetic map. The objective of multipoint linkage analysis is the same as 2 point linkage analysis, but multiple markers are used simultaneously. Since we use more than one marker, we can compute recombination rates more precisely (and order markers more precisely). One account (Lathrop et al., 1985) found the efficiency went up by a factor of 5 by conducting 3 point linkage analysis as opposed to 2 point linkage analysis.

4.6.1 Quantifying linkage

When there is more than one marker, there is more than one way in common use to quantify linkage of a disease locus to the set of markers. We briefly outline these here. All of these measures are log base 10 likelihood ratios under 2 different models (in analogy with the LOD score).

1. **Global support**

   Here we compare the likelihoods when a certain locus is in the map to when it is not on the map. We say there is linkage with the map when the global support exceeds 3.

2. **Interval support**

   Here we compare the likelihood when a locus is in a certain interval to the likelihood when it is in any other interval.

3. **Support for a given order**

   Here we compare the likelihood given a certain order to the most likely order.

4. **Generalized LOD score**

   Here we compare the likelihood given a certain order to the likelihood we get assuming no linkage. For example, if there are 3 loci, we compute

   $$\frac{L(\theta_{12}, \theta_{23})}{L(0.5, 0.5)}.$$

   This measure doesn't make sense when you are mapping a loci onto a known map of markers (because then the denominator doesn't make sense).

5. **Map specific multipoint LOD score**

   Here we express the ratio of the likelihood as a function of locus location to the likelihood when the locus is off the map. A related quantity is the *location score*. This measure uses the usual 2 times the natural logarithm of this likelihood ratio.

To make the idea of the location score more precise, let y_M represent the marker data and y_T the trait data. An important concept for computing likelihoods for pedigrees is the meiosis indicator, $s_{i,j}$, for meiosis i at locus j. We define $s_{i,j}$ to be 0 if the allele comes from the maternal source and 1 if from the paternal source. Note that in general many of the s_{ij} will be only known with uncertainty since we can not determine phase based on the results of genotyping a set of markers. Let s_M be the meiosis indicators for the markers and s_T be those at the putative trait. Suppose the marker map (i.e. the relative location of the markers), Γ_M is known so that $p(s_M)$ is known and the marker allele frequencies determining $p(y_M|s_M)$ are also fixed. Further suppose we know $p(y_T|s_T)$, although this can be parameterized via some parameter vector β which we assume is known. Here $p(y_T|s_T)$ describes some model for how the putative trait locus impacts the trait. For example, we may suppose the trait is dominant, but we will consider other possibilities in Chapter 5. Finally, let γ

Table 4.4 *Frequencies of pairs of recombinations between 3 loci denoted A, B, and C. The letter R indicates a recombination while NR indicates no recombination.*

	BC NR	BC R
AB NR	n_1	n_2
AB R	n_3	n_4

be the location of the trait gene. While γ is unknown, we can vary it and compute

$$p(\gamma|y_T, y_M) \quad \propto \quad p(y_T, y_M|\Gamma_M, \beta, \gamma)$$

$$\propto \quad \sum_{s_M, s_T} p(y_T, y_M, s_M, s_T|\Gamma_M, \beta, \gamma).$$

Then note that

$$p(y_T, y_M, s_M, s_T|\Gamma_M, \beta, \gamma) = p(y_M|s_M)p(s_M|\Gamma_M)p(y_T|s_T, \beta)p(s_T|s_M, \gamma)$$

so only last term depends on γ and the other factors can be ignored if we are just concerned with the most likely location of the trait locus.

4.6.2 An example of multipoint computations

Consider the goal of estimating the relative positions of a set of loci using the generalized LOD score. We will suppose that we can simply count the number of recombinants in our data set. If there are 3 loci (denoted A, B, and C) then there are 4 mutually exclusive possibilities, and these can be assembled as in Table 4.4 (where R stands for a recombination between 2 loci and NR stands for no recombination).

If the 3 loci are all unlinked (i.e. on separate chromosomes) then each of these 4 possibilities is equally likely, whereas if there is linkage between any of the loci, then we expect certain cells of the table to have more entries than other cells. For example, if loci A and loci B where tightly linked we would not expect many entries in the cells labeled n_3 and n_4. To compute the likelihood resulting from such data, we can use the multinomial likelihood. If we let π_{0i} represent the probability of falling in the ith cell in Table 4.4 under the null hypothesis of no linkage (and π_{Ai} is the same probability but under the alternative hypothesis), then the log likelihood ratio is

$$\log \prod_{i=1}^{4} (\pi_{Ai}/\pi_{0i})^{n_i} \quad = \quad \sum_{i=1}^{4} n_i \log(\pi_{Ai}/\pi_{0i})$$

$$= \quad \sum_{i=1}^{4} n_i \log(4\pi_{Ai})$$

since $\pi_{0i} = \frac{1}{4}$ for all i. Under the null all recombination fractions are 0.5, but under the alternative this is not so, hence the next goal is to relate the recombination fractions to the π_{Ai}. But this is simple since for any hypothesis about the relative location of the markers we must have

$$\theta_{AB} = \pi_3 + \pi_4$$
$$\theta_{BC} = \pi_2 + \pi_4$$
$$\theta_{AC} = \pi_2 + \pi_3$$

which then implies

$$\pi_1 = (\theta_{AB} - \theta_{AC} + \theta_{BC})/2$$
$$\pi_2 = (\theta_{AC} - \theta_{BC} + \theta_{AB})/2$$
$$\pi_3 = (\theta_{BC} - \theta_{AB} + \theta_{AC})/2$$
$$\pi_4 = 1 - \pi_2 - \pi_3 - \pi_1,$$

hence we obtain an expression for the generalized LOD score in terms of the recombination fractions by using these expressions in the likelihood (e.g. setting $\pi_{A1} = (\theta_{AB} - \theta_{AC} + \theta_{BC})/2$). We can then maximize this expression and obtain MLEs of the recombination fractions. In contrast, if we conducted 3 2 point analyses, the MLEs of θ_{AB}, θ_{BC} and θ_{AC} are $(n_3 + n_4)/\sum_i n_i$, $(n_2 + n_4)/\sum_i n_i$ and $(n_2 + n_3)/\sum_i n_i$ respectively. Intuitively, if loci A and B are close, then data about the recombination fraction between A and C is relevant for estimating the recombination fraction between B and C, hence multipoint analyses are potentially more precise than 2 point methods.

4.7 Computation of pedigree likelihoods

As the simple pedigree examples indicate, computing the likelihood for a pedigree can be difficult due to the fact that we don't know phase. There are a variety of algorithms for calculating the likelihood for general pedigrees. The appropriate method depends on the problem at hand. The first general method for calculating the likelihood for general simple pedigrees was devised by Elston and Stewart (1971). The idea of this method is to realize given the genotype information, the phenotypes of distinct individuals are independent (since they only depend on genotypes). In addition, given the parents' genotype information, the grandparents' genotypes don't help predict the childrens' genotype: that is, the childrens' genotypes are conditionally independent of the grandparents' genotypes, given the parents genotypes. Since distinct families are independent, one just considers computing the likelihood for one family (the total likelihood is the product of these factors). An alternative is the Lander Green algorithm (1987). The idea behind this algorithm is that as one moves along a chromosome we can think of the data as arising from a stochastic process. If we assume this follows a first order Markov chain then we can decompose the likelihood by factoring the information from markers across the chromosome. This algorithm's complexity increases linearly with the number of loci considered and exponentially in pedigree size, whereas the Elston Stewart algorithm has a computational complexity that increases linearly with pedigree size and exponentially in the number of loci considered.

4.7.1 The Elston Stewart algorithm

The Elston Stewart algorithm computes the likelihood by utilizing the assumption that conditional on the genotypes, all phenotypes are independent. Note that this isn't necessarily true since environment will also impact phenotypes and this will usually be largely common among members of the same family.

Let y represent the vector of phenotypic data, and suppose $y_{i,.}$ is the phenotypic data for subject i. If g is set of all genotypes, and g_i is the genotype for subject i, then the Elston Stewart algorithm is based on the decomposition

$$
\begin{aligned}
L(y) &= \sum_g p(y|g)p(g) \\
&= \sum_g \left(\prod_i p(y_{i,.}|g_i)\right)p(g).
\end{aligned}
$$

Next note that if the pedigree is simple and we let $\mathcal{G}_t$ represent the set of subjects in generation t for $t = 1, \ldots T$, we can write

$$
p(g) = p(g_i \text{ for } i \in \mathcal{G}_1) \prod_{t=2}^T p(g_i \text{ for } i \in \mathcal{G}_t | g_i \text{ for } i \in \mathcal{G}_{t-1}).
$$

We usually assume that generation 1 corresponds to the founders, although this is not necessary. Given a model for the genetic mode of transmission it is usually straightforward to compute $p(g_i \text{ for } i \in \mathcal{G}_t | g_i \text{ for } i \in \mathcal{G}_{t-1})$. Note that computing this term will involve if a recombination has occurred between the tth and $t - 1^{\text{th}}$ generations, hence these conditional probabilities will depend on θ. While this algorithm has been used extensively, it is limited by the size of the sum over all possible genotypes. The computational burden will grow exponentially with the number of markers used due to this sum, and this can make the algorithm impractical for some data sets. The program LINKAGE can be used to carry out the computations necessary for the Elston Stewart algorithm.

4.7.2 The Lander Green algorithm

Similarly to the decomposition used in the Elston Stewart algorithm, if s is the set of all meiosis indicators then we can also write

$$
L(y) = \sum_s p(y|s)p(s).
$$

Hence, if we let $y_{.,j}$ represent the data for the jth marker, $s_{.,j}$ to represent the meiosis indicators at the jth marker and $s_{i,.}$ to represent the meiosis indicators for subject i then if the data are for known loci and are ordered along the chromosome in j we have (since meioses are independent across matings)

$$L(y) = \sum_s \left(\prod_j p(y_{.,j}|s_{.,j}) \right) \left(\prod_i p(s_{i,.}) \right).$$

Note that recombination parameter only enters $\prod_i p(s_{i,.})$ and parameters for the genetic model for the trait and marker loci enter the other factor.

When discussing multilocus mapping, the impact of recombinations in one location on recombinations in nearby regions of the genome must be considered. It is possible that a recombination at one location may make recombinations at nearby locations less likely. In such a case we say there is interference since one recombination interferes with the occurrence of others. Usually markers in linkage studies are widely separated on the genome so that interference is unlikely to be a major issue. If there is no interference then the inheritance vector, $s_{.,j}$, has the Markov property as you move across the chromosome, so, if there are a total of L loci under consideration, we can write

$$L(y) = \sum_s p(s_{.,1}) \prod_{j=2}^{L} p(s_{.,j}|s_{.,j-1}) \prod_{j=1}^{L} p(y_{.,j}|s_{.,j}).$$

Then we can treat the values of $s_{.,j}$ as a hidden Markov process (discussed more fully in Chapter 9) and use the Baum-Welch algorithm (which is a special case of the EM algorithm for hidden Markov models) to maximize the likelihood in a way that is linear in the number of markers but exponential in the number of meioses. The programs GENEHUNTER and ALLEGRO implement the Lander Green algorithm and are widely available.

4.7.3 MCMC based approaches

Note the similarity between the Lander Green and Elston Stewart algorithms: both do something like

$$L(y) = \sum_x p(y|x)p(x)$$

where x is either the set of genotypes (for the Elston Stewart algorithm) or the set of meiosis indicators (for the Lander Green algorithm). In general we won't know the values of the parameters that enter into the likelihood, and so must try to estimate these. The basic idea is to use MCMC combined with the above algorithms. To do so, one treats x as a latent variable and samples this, then given these, one can use the Gibbs sampler to sample the model parameters (such as the recombination fraction). Once the parameters that enter the genetic model have been sampled, one then samples either g or s. If one samples loci, i.e. g, then the algorithm is referred to as an L-sampler. If one instead samples the meioses then the algorithm is referred to as an

M-sampler. Current approaches combine the 2 types of moves. The software LOKI can be used to conduct these computations (see Shmulewitz et al., 2006 for further details).

4.7.4 Sparse binary tree based approaches

While MCMC methods have allowed for consideration of large pedigrees, for sets of markers that are tightly linked, these methods can have difficulty in that the Markov chains converge slowly to the limiting distribution. Thus another method that is commonly used to compute likelihoods for pedigrees uses binary trees to speed up the computational time of the Lander Green algorithm. The main difficulty for the Lander Green algorithm is that the algorithm computes the likelihood for every possible inheritance vector. Since each element of the inheritance vector is either 0 or 1 and given that the length of the inheritance vector depends on the number of meioses, the number of possible inheritance vectors increases exponentially with the number of meioses. However, the likelihood is exactly the same for many of these inheritance vectors. The idea behind the use of sparse binary trees is to replace each element of the inheritance vector with a node in a binary tree (so each node has 2 leaves). The 2 leaves of the tree correspond to transmission of the maternal or paternal allele to an offspring. When all further nodes in some part of the tree lead to the same value for the likelihood, the rest of the tree beyond that node isn't considered, hence the tree ends at that location. This allows one to not have to consider the many inheritance vectors that lead to the same value of the likelihood, and this greatly speeds up the computational time, up to approximately a factor of 10 (Abecasis et al., 2002). The software program called MERLIN implements this approach.

4.8 Exercises

1. Show that the heterozygosity, H, is maximized when all alleles are present in equal proportions in the population.

2. Show that $\frac{n}{n-1}\hat{H}$ is an unbiased estimate of the heterozygosity by using the mean and variance of a sample proportion.

3. In this exercise we will compute an approximation to the number of markers necessary to localize a gene to a given resolution with a certain probability.

   (a) Given a map of length L cM, what is the probability of an arbitrary locus being no more than d cM from a randomly located marker (*Hint*: treat the genome as a string of length L, suppose there is a fixed location for some marker, and suppose the locus is uniformly distributed over the genome)?

   (b) Compute the probability, p, that a gene is within distance d of at least one of m markers (*Hint*: use the approximation that the event that a gene is within distance d of marker i is independent of the event that a gene is within distance d of marker j for all i, j).

   (c) Using your expression from part (b), express m as a function of p, d and L.

(d) For humans, averaging over the sexes, $L = 3300$ cM. How many markers do you need to localize a gene if you want to ensure we stand a 95% chance of having a marker within 10 cM of a given gene?

4. Use the method of Smith (1959) to compute the posterior probability of linkage when you observe 2 recombinants out of 10.

5. Suppose that in n trials 0 recombinations are observed. Find a 95% confidence interval for the recombination fraction.

6. If one has data from a nuclear family where all family members are doubly heterozygous (i.e. all subjects have genotype (1,2) at 2 loci with unknown phase), is the family informative in a linkage study? Explain.

7. For X-linked traits, researchers often use a significance level of $\alpha = 0.002$, what value does the LOD score need to reach to obtain this level of significance?

8. Show how to find confidence intervals for the recombination fractions when considering 3 loci where the number of recombinations between all loci is observed in a multipoint context.

9. Compare the variance estimates of the recombination fraction in a 3 point multipoint linkage analysis compared to a 2 point analysis. Suppose phase is known for all subjects.

CHAPTER 5

Extensions of the Basic Model for Parametric Linkage

5.1 Introduction

Our model regarding the relationship between genotypes and phenotypes thus far has been overly simplistic. We have assumed that each genotype translates into one definite phenotype, and while this is a useful model for studying some genetic disorders, it is not realistic for many other disorders. A good example of where this simple model breaks down is with breast cancer. At least one gene has been found where the presence of certain alleles seems to be associated with developing breast cancer. The relationship here is complicated since not everyone who has this gene develops breast cancer by a certain age, hence we speak of *susceptibility* genes rather than disease genes. Taking this perspective greatly extends the applicability of finding genes associated with disorders.

In addition, we have assumed all genetic disorders fit into the Mendelian framework of a single disease gene (that is either dominant or recessive), but there is increasing evidence that other modes of transmission provide better models for inheritance. Recall that a trait whose mode of genetic transmission does not fit into the Mendelian framework is referred to as a complex trait. Additionally, it is possible that genetic transmission of a trait is well modeled by the Mendelian framework at more than one locus, and different affected individuals have genetic anomalies at different loci.

By moving outside of the Mendelian framework, a vast array of models for disease etiology is possible. Unfortunately it is not always possible to determine if a given model is identified. All that is observed is phenotypes, hence if we relax the manner in which genotypes of parents are transformed into genotypes of offspring and the manner in which genotypes are transformed into phenotypes we run the risk of using nonidentifiable models (since we are proposing 2 models for unobservable phenomena). This risk is very serious because computing the likelihood for pedigree data is not transparent, hence it is difficult to judge if a given model is identifiable. This implies that a nonlinear maximization routine may converge to a local mode when in fact the likelihood is unbounded or no unique MLE exists (because the model is unidentifiable).

5.2 Penetrance

Penetrance is defined as the probability one has a certain phenotype given the genotype. Thus far, we have assumed penetrance is either zero or one. In general, we can allow the penetrance to be a parameter and maximize the likelihood with respect to the recombination fraction and the penetrance simultaneously. One problem with this approach is *ascertainment bias*, that is, the families we have in a typical genetic study are selected because some member of the family has the disease and the genotype. We miss individuals who have the genotype but not the phenotype, hence other methods for estimating penetrance have been devised. Clearly, ignoring the ascertainment bias will result in an overestimate of the penetrance.

If we are not interested in the recombination fraction, and we suppose the disease is a rare, dominant disorder, then we can estimate the penetrance with $2\hat{p}$, where $\hat{p}$ is the proportion of affecteds in the family. This is because

$$
\begin{aligned}
P(\text{affected}|\text{genotype}) &= \frac{P(\text{genotype}|\text{affected})P(\text{affected})}{P(\text{genotype})} \\
&= \frac{P(\text{affected})}{P(\text{genotype})} \\
&= \frac{P(\text{affected})}{1/2}
\end{aligned}
$$

since $1/2$ of the offspring from matings in rare dominant families would be affected on average and $P(\text{genotype}|\text{affected}) = 1$. Similarly, for a rare, recessive trait we would use $4\hat{p}$. As above, this estimate will be biased due to the ascertainment bias. *Weinberg's proband method* (Weinberg, 1927) attempts to correct for ascertainment bias by leaving the proband out of the calculation, that is, if y_i represents the number of affected individuals in the ith kinship (of size n_i), then

$$
\hat{p} = \frac{\sum_i (y_i - 1)}{\sum_i (n_i - 1)}.
$$

Some disorders only develop with age, hence we speak of *age dependent penetrance*. Instead of just estimating a single parameter representing the penetrance, we parameterize the probability of having a certain phenotype by age y given a certain genotype. For example, if y represents the age at which an individual with a certain genotype develops some disease, we might use the logistic curve to model the penetrance for a given genotype

$$
P(y \leq t|\mu, \sigma) = \frac{1}{1 + \exp\{-(y - \mu)/\sigma\}},
$$

and estimate the parameters using maximum likelihood or Bayesian methods (see, for example, Gauderman and Thomas, 1994). In an analogous fashion, we can allow the penetrance to depend on any characteristic we desire.

5.3 Phenocopies

A *phenocopy* is an individual who displays a disease phenotype but does not have the disease genotype. It is often reasonable to suppose that a certain trait can have environmental sources and genetic sources, and some individuals have the disorder due to environmental sources alone. The *phenocopy rate* is defined as the probability that one is a phenocopy given that he or she is affected. Clearly a high phenocopy rate will make it difficult to establish linkage as only a subset of affecteds will actually have linkage of a marker to the disease locus. The best means for getting around this issue is careful definition of the phenotype. Many disorders have been defined with reference to clinical criteria (since medical management is the usual objective), but clinical criteria often lump cases into categories which may not be useful from the perspective of determining the genetic source of the disorder. For example, obesity is a harmful condition, and from the perspective of clinical management, it may not make much of a difference as to the source of the obesity. Yet there seem to be genetic cases of obesity, and if the distinction between genetic and non-genetic cases is not made it will be very difficult to find any obesity susceptibility genes. The solution is to try to use a phenotype definition that distinguishes between the genetic and non-genetic cases, such as some aspect of the metabolism, rather than, say, body mass index.

5.4 Heterogeneity in the recombination fraction

It is well documented that there is heterogeneity in the recombination fraction. For example, there are more crossovers in women than in men (as documented in Li, Fann and Ott, 1998), hence probability models should have a recombination fraction for men and women. In *D. melanogaster* the males show no recombination at all. Moreover, there appears to be heterogeneity in the recombination fraction between families for some sets of loci. Since normal physiology is the outcome of many inter-related biological pathways, there are many opportunities for something to go wrong. For this reason, distinct genetic defects can have the same outcome in terms of physiology, that is, distinct genotypes can lead to the same disease phenotype. Often this sort of heterogeneity is referred to as *locus heterogeneity* in contrast to *allelic heterogeneity*. Allelic heterogeneity is the situation in which more than one allele at the same locus can lead to the disorder. Cystic fibrosis is an example of a disorder characterized by allelic heterogeneity.

If heterogeneity is ignored, it will be difficult to establish linkage, or it may appear that several genes are involved in the disorder (leading to the incorrect conclusion that the disorder is complex). For this reason, a number of statistical tests have been devised to test for heterogeneity and linkage in the presence of heterogeneity. With large families, a sensible first step is to conduct a linkage analysis for each family separately and all families together. Additionally, one should examine all the phenotypic data to see if heterogeneity is present with regard to the phenotype.

5.4.1 Heterogeneity tests

Suppose the families can be grouped into c classes based on the phenotype data (for example, ethnicity). Alternatively, there may be c families. Morton (1956) proposed to test the null hypothesis

$$H_0 : \theta_1 = \cdots = \theta_c < \frac{1}{2}$$

against the alternative

$$H_1 : \theta_1 \neq \cdots \neq \theta_c.$$

If $\hat{\theta}_i$ is the MLE for the ith class, $\hat{\theta}$ the MLE ignoring the classes, Z_i is the log 10 likelihood ratio for the ith class, and Z is log 10 likelihood ratio ignoring the classes, then Morton proposed the following maximum likelihood ratio test

$$X^2 = 2\log(10) \left[\sum_i Z_i(\hat{\theta}_i) - Z(\hat{\theta}) \right].$$

Under H_0, X^2 is asymptotically χ^2 with $c - 1$ degrees of freedom (since we can obtain H_1 from H_0 by imposing $c - 1$ linear constraints). It is possible to use this test and reject homogeneity, yet fail to establish linkage (especially since conventional significance levels are often used when conducting this test), hence one suggestion (Ott, 1999) is to use a significance level for this test of 0.0001.

Other methods don't require one to break up the families into classes, or suppose the families are all in their own class. The basic idea is to suppose there is some distribution from which each family has a draw that determines the recombination fraction for that family, that is, we use a *hierarchical model*. The simplest example is to suppose there are 2 family types: those with linkage of the disease trait to a marker and those without linkage. In this case, if we let α represent the probability that a given family has linkage of the disease gene to the marker, the likelihood for the ith family can be expressed

$$L_i(y|\alpha, \theta) = \alpha L_i(y|\theta) + (1 - \alpha)L_i(y|\theta = 0.5).$$

We then test the hypothesis $H_0 : \alpha = 1$ against the alternative $H_1 : \alpha < 1, \theta < 1/2$. If we define the test statistic as

$$X^2 = 2 \left[\log L(y|\hat{\alpha}, \hat{\theta}) - \log L(y|1, \hat{\theta}) \right],$$

then X^2 has a distribution under the null hypothesis that is a χ^2 variable with one degree of freedom with probability 0.5, and is zero with probability with 0.5. To get a p-value, look up the value for X^2 in a χ^2 table (with 1 degree of freedom), then divide that probability by 2 (Ott, 1983). There are other tests for admixture based on more sophisticated hierarchical models. One can incorporate a variety of factors into such models, such as age of onset.

Rather than testing for heterogeneity, we may want to test for linkage given there is heterogeneity. In the framework of the test for admixture, we are now interested in the test statistic

$$X^2 = 2 \left[\log L(y|\hat{\alpha}, \hat{\theta}) - \log L(y|\hat{\alpha}, 0.5) \right].$$

The distribution of this test statistic is well approximated as the maximum of 2 independent χ^2 variables with 1 degree of freedom. If we take the log to the base 10 of the maximized likelihood ratio, then we get the *hLOD* (Elmsie et al., 1997).

5.5 Relating genetic maps to physical maps

Genetic distance is directly related to physical distance. As either distance grows, the genetic distance departs from the physical distance since genetic distance is between 0 and 0.5 but physical distance is only bounded by the length of a chromosome. To understand the nature of the problem, realize that if one typed 3 markers on a single chromosome and assumed that genetic distances add in the same manner that physical distance on a line is additive, then one could conclude that the recombination fraction between 2 of the markers exceeds 0.5. Note that if we have a model for the mapping from physical distance to genetic distance, denoted $M(x)$, we can use linkage based approaches to do physical mapping (or vice versa) since if $\hat{\theta}$ is the MLE of the genetic distance the MLE of the physical distance will just be $M^{-1}(\hat{\theta})$.

Consider 3 loci, A, B, and C, and denote the recombination fractions as θ_{AB}, θ_{AC} and θ_{BC}. Suppose the order of these loci along the chromosome is given by A, B, C. We will denote the event that there is a recombination between loci A and B as AB while if no recombination occurs, we will denote that event as $\overline{AB}$ (with similar notation for a recombination between other pairs of loci). Then we have the following

$$
\begin{aligned}
\theta_{AC} &= P((AB \text{ and } \overline{BC}) \text{ or } (BC \text{ and } \overline{AB})) \\
&= P(AB \text{ and } \overline{BC}) + P(BC \text{ and } \overline{AB})
\end{aligned}
$$

so that, by the definition of conditional probability,

$$
\begin{aligned}
\theta_{AC} &= \theta_{AB}P(\overline{BC}|AB) + \theta_{BC}P(\overline{AB}|BC) \\
&= \theta_{AB}(1 - P(BC|AB)) + \theta_{BC}(1 - P(AB|BC)) \\
&= \theta_{AB} + \theta_{BC} - \theta_{AB}P(BC|AB) - \theta_{BC}P(AB|BC) \\
&= \theta_{AB} + \theta_{BC} - \theta_{AB}\theta_{BC}\left(\frac{P(BC|AB)}{\theta_{BC}} + \frac{P(AB|BC)}{\theta_{AB}}\right).
\end{aligned}
$$

Next, by Bayes theorem

$$
P(AB|BC) = \frac{P(BC|AB)\theta_{AB}}{\theta_{BC}},
$$

so that

$$
\begin{aligned}
\theta_{AC} &= \theta_{AB} + \theta_{BC} - \theta_{AB}\theta_{BC}P(BC|AB)\left(\frac{1}{\theta_{BC}} + \frac{\theta_{AB}}{\theta_{AB}\theta_{BC}}\right) \\
&= \theta_{AB} + \theta_{BC} - 2\theta_{AB}P(BC|AB).
\end{aligned}
$$

If we then define $\gamma = \frac{P(BC|AB)}{\theta_{BC}}$ we then find that

$$
\theta_{AC} = \theta_{AB} + \theta_{BC} - 2\theta_{AB}\theta_{BC}\gamma.
$$

The parameter γ is referred to as the *coefficient of coincidence*. A related quantity, known as the *interference* is defined as $I = 1 - \gamma$. We expect γ to be less than one for sets of close loci due to it being unlikely that there are multiple crossovers over a short distance. One can estimate γ by using the estimated recombination fractions, but usually estimates are very imprecise. It is typically assumed that γ does not depend on genomic location, but only the physical distances between the loci involved.

If γ is known, then we can relate physical distance to recombination probabilities. A method due to Haldane (1919) for understanding the relationship between map distance and recombination fractions supposes the relationship between map distance is given by the identity map for small distances (in some units for physical mapping, such as base-pairs). First, given the above definition of γ, we can write

$$\gamma = \frac{1}{2} \frac{\theta_{AB} + \theta_{BC} - \theta_{AC}}{\theta_{AB}\theta_{BC}}.$$

Now suppose the physical distance between A and B is x, and suppose C is close to B, and of distance Δx. Using $\theta = M(x)$ to denote the mapping of physical distances into genetic distances, then, if $M(0) = 0$, we have the following

$$\begin{aligned} \gamma &= \frac{1}{2} \frac{M(x) + M(\Delta x) - M(x + \Delta x)}{M(x)M(\Delta x)} \\ &= \frac{1}{2} \frac{1}{M(x)} \left(\frac{M(\Delta x)}{M(\Delta x)} - \frac{M(x + \Delta x) - M(x)}{M(\Delta x)} \right) \\ &= \frac{1}{2} \frac{1}{M(x)} \left(1 - \frac{\big(M(x + \Delta x) - M(x)\big)/\Delta x}{\big(M(\Delta x) - M(0)\big)/\Delta x} \right) \end{aligned}$$

then taking Δx to zero we find

$$\gamma = \frac{1}{2M} \left(1 - \frac{M'(x)}{M'(0)} \right).$$

If the map is linear with slope one for small x, i.e. $M'(0) = 1$, then we obtain the differential equation

$$\gamma_0 = \frac{1}{2M} \left(1 - M'(x) \right).$$

We use γ_0 here to recognize the role of the assumptions $M(0) = 0$ and $M'(0) = 1$, and we call γ_0 *Haldane's marginal coincidence coefficient*. Hence we have

$$\frac{dM}{dx} = 1 - 2M\gamma_0.$$

Since $M(0) = 0$ we can solve this equation once we specify γ_0. For example, if $\gamma_0 = 1$, $M(x) = (1 - e^{-2x})/2$, a map function known as *Haldane's map function*. The assumption $\gamma = 1$ is interpreted as $P(BC|AB) = P(BC)$, i.e. independence of crossovers (a case known as no interference). Recall the Lander Green algorithm assumes no interference. If $\gamma_0 = 0$ then there are no crossovers close to a known crossovers, a case known as complete interference (see Exercise 5 for more on this case). If $\gamma_0 = 2\theta$, then $M(x) = \tanh(2x)/2$, which is *Kosambi's map function*.

Many other map functions in the literature can be viewed as arising in this way, and since we can interpret γ_0, we can interpret the meaning of a map function in terms of the conditional probability of a crossover given a nearby crossover.

Given a map function, one can compute recombination fractions between 2 loci, A and C, given the recombination fractions between each locus and some intermediate locus. The resulting formulas are known as *addition rules*, and depend on the form of the map function. If x_{AB} is the physical distance between loci A and B and C is another loci so that B resides between A and B, then since physical distance along the chromosome is additive we have (in obvious notation)

$$x_{AC} = x_{AB} + x_{BC}.$$

Hence given a map function M so that $M(x_{AB}) = \theta_{AB}$, if M is strictly monotone we find that $x_{AB} = M^{-1}(\theta_{AB})$ so that

$$M^{-1}(\theta_{AC}) = M^{-1}(\theta_{AB}) + M^{-1}(\theta_{BC}),$$

which implies that

$$\theta_{AC} = M\left(M^{-1}(\theta_{AB}) + M^{-1}(\theta_{BC})\right).$$

This equation gives an explicit form for the addition rule given some map function.

One problem with map functions generated by Haldane's method is they do not necessarily give rise to valid probabilities for observing gametes, that is, they are not necessarily *multi-locus feasible*. If there are more than 3 loci under consideration, one can construct examples where the probability of observing certain recombination patterns is less than zero (see Liu, 1998 for an example). For a full treatment, one needs to determine what is the probability of observing certain recombination patterns. For example, if there is no interference one can obtain the probability of observing gametes by just multiplying probabilities of recombinations between distinct loci. Suppose there are 3 loci and someone is heterozygous at all loci. Further, use letters to refer to different loci and distinguish between alleles via case. Then with no interference, the probability of observing a gamete with alleles A, B, c is just $\frac{1}{2}(1 - \theta_{AB})\theta_{BC}$ since there is no recombination between loci A and B however there is between loci B and C, and given that this event has taken place the 2 chromosomes A, B, c and a, b, C are equally likely. When there is interference, computing the gamete probabilities is more complex, but expressions are available to find these (see Liberman and Karlin, 1984).

Liberman and Karlin (1984) show that a map is multi-locus feasible if

$$(-1)^k M^{(k)}(x) \le 0,$$

for all $k \ge 1$ and all $x \ge 0$. For example, Kosambi's map is not multi-locus feasible. A way to construct maps that are multi-locus feasible is to fix the maximum number of crossovers, but let the locations be random. If $f(s)$ is the probability generating function for the number of crossovers, i.e.

$$f(s) = \sum_k p_k s^k,$$

where p_k is the probability of k crossovers, then, if μ denotes the mean number of crossovers, we define the map

$$M(x) = \frac{1}{2}\left[1 - f\left(1 - \frac{2x}{\mu}\right)\right]$$

As an example, suppose the number of crossovers is distributed according to the binomial distribution with parameters n and p. Then

$$
\begin{aligned}
f(s) &= \sum_{k=1}^{n} \binom{n}{k} p^k (1-p)^{n-k} s^k \\
&= \sum_{k=1}^{n} \binom{n}{k} (sp)^k (1-p)^{n-k} \\
&= (ps + 1 - p)^n
\end{aligned}
$$

by the binomial theorem, so

$$f\left(1 - \frac{2x}{\mu}\right) = \left(1 - \frac{2x}{n}\right)^n,$$

hence

$$M(x) = \frac{1}{2}\left[1 - \left(1 - \frac{2x}{n}\right)^n\right]$$

If $n = 1$ we find $M(x) = x$, while if we take the limit in n we obtain Haldane's map. This makes sense because if $n = 1$ then there is only 1 possible crossover, hence if this occurs no more crossovers can take place, i.e. there is complete interference. In contrast, as $n \to \infty$ the occurrence of a crossover has no effect on the occurrence of another.

5.6 Multilocus models

Thus far we have only assumed that a single locus impacts a phenotype, however this is unlikely for complex traits. Many biological systems have backup systems, and this can lead to the appearance of the interaction between 2 genes. For example, suppose that there are 2 ways a cell can bring about some desired outcome (through 2 distinct biological pathways). If this is the case, then provided either of the pathways are intact, the phenotype will appear normal. However, if both pathways are negatively affected by disease mutations, then the cell would be unable to obtain the desired outcome, and the phenotype would change. In this case there would appear to be an interaction between pairs of genes in the distinct pathways. Given the general formulation of the likelihood for a pedigree, one can construct models that have multiple loci all of which contribute to the risk of developing disease. However, most analyses just look at the effect of single genes since most argue that the effects of each locus should be substantial if there are multiple loci involved.

5.7 Exercises

1. Show that if the prevalence of a disease genotype is less than the prevalence of the associated disease phenotype then the penetrance will exceed the phenocopy rate.

2. Consider a family with 2 parents and 1 child. Suppose one parent and the child have a rare, dominant genetic disorder. We genotype the family at one marker locus and find the affected parent has genotype (1,2), while the other parent and the child have genotype (2,2). Compute the likelihood assuming incomplete penetrance as a function of the recombination fraction, genotype frequencies and the penetrance.

3. Show that Morton's test is distributed as a χ^2_{c-1} random variable.

4. Describe how to compute the p-value for the hlod.

5. Morgan's map functionn can be found by supposing $\gamma_0 = 0$. Use Haldane's method to find the map, $M(x)$, to map physical distance into genetic distance for this marginal coincidence coefficient.

6. If we assume complete interference, explain why all chromosomes have the same length in terms of genetic distance. What is this common length?

7. Derive the addition formula for the Kosambi map function.

8. Show that Kosambi's map function is not multilocus feasible.

CHAPTER 6

Nonparametric Linkage and Association Analysis

6.1 Introduction

In the context of linkage analysis, a nonparametric method is a technique for establishing linkage that does not require specification of the mode of transmission. This is in contrast to the previous methods in which we had to specify if the disease was recessive or dominant (or perhaps the penetrance was specified). A common feature of nonparametric methods is the use of small, nuclear families as opposed to the large multigeneration families used in parametric linkage analysis.

6.2 Sib-pair method

In the 1930s and 1940s, the most popular method of linkage analysis was based on the sib-pair method of Penrose (1935). While subsequent developments have supplanted this method, it is provides a natural introduction to modern affected sib-pair methods. In addition, this method uncovered the first case of linkage in humans.

The method is quite simple. Suppose the phenotype for each member of a sib-pair (i.e. a pair of siblings) is known. We suppose each phenotype corresponds to a genotype at some locus. Each sib-pair can be classified as having the same or not the same phenotype for each of the 2 phenotypes. We organize this data into a 2 by 2 table, as in Table 6.1, with a dimension for each phenotype, and a row or column for same or different. Each sib-pair contributes one entry to this table.

Table 6.1 *Relationship between 2 phenotypes for a set of pairs of siblings.*

	same on phen. 2	differ on phen. 2
same on phen. 1	n_1	n_2
differ on phen. 1	n_3	n_4

If there is linkage between the loci corresponding to the phenotypes, the concordant cells of the table should fill up, i.e. sib-pairs will either have the same allele at the 2 loci or different alleles. We can then conduct a goodness of fit test, such as Pearson's χ^2 test, to test for linkage.

While this method is simple to implement and easy to understand, it has some weaknesses. First, since it ignores the parental phenotypes, potentially many non-informative matings are considered in the data, and this acts to dilute any effect we may otherwise see. Secondly, incomplete penetrance and phenocopies can disguise effects since we don't account for these factors.

6.3 Identity by descent

To discuss the usual treatment of quantitative traits we need to introduce the idea of 2 markers being *identical by descent* (IBD). Consider 2 offspring and the alleles they have at some locus. If the 2 received the allele from the same parent, then the alleles are said to be identical by descent. If they have the same value for an allele but didn't receive it from the same parent, we say the alleles are *identical by state* (IBS). As an example, suppose the parents have 4 different marker alleles at some locus, denoted A_1, A_2 for one parent and A_3, A_4 for the other. Two of their children will either have 0, 1, or 2 alleles in common at this locus. For, suppose that the first parent transmitted the A_1 allele to the first child, and a A_2 allele to the second child while the other parent transmitted a A_3 allele to the first child and a A_4 allele to the second child. Then the genotypes of the offspring would be (A_1, A_3) and (A_2, A_4), and at this locus the 2 offspring would have no alleles in common. On the other hand, if the first parent transmitted the A_1 allele to both offspring and the second parent transmitted the A_3 allele to both offspring then both offspring would have the genotype (A_1, A_3) and would share 2 alleles identical by descent. If the marker locus is unlinked to the disease locus and both children are affected, the probability they share 0 or 2 alleles is 1/4 and the probability they share 1 allele is 1/2. On the other hand, if the marker is tightly linked to the disease locus, then you would expect children to share alleles at the marker locus more frequently than not (with the surplus depending on the mode of genetic transmission). For example, suppose a marker locus is tightly linked (i.e. the recombination rate is zero) to a recessive trait and neither parent is affected while both offspring are. Then the 2 offspring must each have received 1 copy of the disease allele from each parent. But then the 2 offspring must have received the allele at the marker that was on the chromosome with the disease allele from each parent and so the 2 offspring would have 2 alleles IBD at the marker locus. Frequently IBD status is treated as a percentage so that if 2 siblings share 2 alleles IBD we say that their IBD status is 1.

6.4 Affected sib-pair (ASP) methods

A simple modification to the idea of the basic sib-pair method greatly enhances the power of the method. Instead of using any sib-pair, we use *affected* sib-pairs (some methods also use affected-unaffected sib-pairs). While we may still consider non-informative matings (since the parents may be homozygous for the marker), we don't have to worry about incomplete penetrance since both of the siblings are affected (although phenocopies still pose a problem). A variety of tests have been designed to test for linkage by comparing IBD proportions to their expected values under the hypothesis of no linkage. To discuss these test, define

$$z_i = P(i \text{ alleles are IBD for an ASP}),$$

so that if a marker is tightly linked to a recessive disease gene where both parents are unaffected, we expect $z_0 = z_1 = 0$ and $z_2 = 1$, for a rare, dominant disorder $z_0 = 0$, and $z_1 = z_2 = 1/2$, and under no linkage $z_0 = 1/4$, $z_1 = 1/2$ and $z_2 = 1/4$.

The basic idea of any test using ASPs is to estimate the set of z_i with their corresponding sample proportions, then construct a test statistic based on these proportions. One could construct a χ^2 goodness of fit test for these proportions (a test with 2 degrees of freedom), but this test will have low power compared to tests that have a specific alternative. If one specifies a mode of transmission in the alternative hypothesis, then one can construct a more powerful test than this χ^2 goodness of fit test. As is typical for non-parametric methods, one can devise a test that doesn't need a specification of the model, but one can devise a more powerful test if one proposes a model under the alternative. For this reason, there are 3 popular methods for testing the hypothesis of no linkage in ASP analysis. One specifies the alternative of a recessive mode of inheritance, one specifies a dominant mode of inheritance, and one tries to find a balance between the other two. Not surprisingly, the test that tries to strike a balance is not as powerful as either of the other 2 tests if the mode of inheritance is known.

6.4.1 Tests for linkage with ASPs

The mean test is designed to provide a powerful test under the alternative of a dominant mode of transmission. The test statistic is

$$T_{\text{mean}} = \hat{z}_2 + \frac{1}{2}\hat{z}_1.$$

One compares this to the null value of 1/2, and the distribution of the statistic can be worked out based on the properties of sample proportions from multinomial samples. That is, the variance of a sample proportion is well known, but here we have to consider the correlation between $\hat{z}_2$ and $\hat{z}_1$.

The proportion test is designed to provide a powerful test when the alternative is a recessive mode of transmission. The test statistic is

$$T_{\text{proportion}} = \hat{z}_2,$$

and one compares this to its null value of 1/4. The variance of this statistic is just the variance of a sample proportion. Recall, for recessive traits we expect $z_0 = z_1$, so we use neither of these in the test.

Whittemore and Tu (1998) propose a test of no linkage using the idea of minimax tests. Both the mean and proportion tests can be viewed as a special case of the more general test statistic

$$T_w = w_0 \hat{z}_0 + w_1 \hat{z}_1 + w_2 \hat{z}_2.$$

Note that we can re-express T_w as

$$
\begin{aligned}
T_w &= w_0(1 - \hat{z}_1 - \hat{z}_2) + w_1 \hat{z}_1 + w_2 \hat{z}_2 \\
&= w_0 + (w_1 - w_0)\hat{z}_1 + (w_2 - w_0)\hat{z}_2 \\
&= w_0 + (w_2 - w_0)\left(\frac{(w_1 - w_0)}{(w_2 - w_0)} \hat{z}_1 + \hat{z}_2 \right).
\end{aligned}
$$

Hence the distribution of the test statistic just depends on what value is selected for $c_1 = (w_1 - w_0)/(w_2 - w_0)$, since other values of w_i just rescale and recenter the test statistic $T_w = c_1 \hat{z}_1 + z_2$. The test statistic $c_1 \hat{z}_1 + z_2$ will be asymptotically normal (since it is a linear combination of sample proportions), so if we use the square of T_w we get a test statistic that is asymptotically χ^2 with 1 degree of freedom (as opposed to the 2 degrees of freedom in the likelihood ratio test). This implies the properties of the test depend solely on the value of c_1. The idea of Whittemore and Tu is to select w_1 so that we minimize the maximal cost of making a mistake. An interesting result of Whittemore and Tu's in this connection is that any allowable value for c_1 must lie between 0 and 1/2 (the values used in the proportion and mean tests respectively). They show that setting $c_1 = 0.275$ is optimal in a certain sense, which is slightly closer to the mean test.

6.5 QTL mapping in human populations

There has recently been considerable interest in mapping genes associated with certain quantitative traits. Given the general form for the likelihood for a pedigree, it is conceptually straightforward to allow for incomplete penetrance and a continuous density for the trait values (such as normal or log-normal). While this method allows one to model quantitative traits and thereby find loci that control quantitative traits (refered to as quantitative trait locus (QTL)), there are a number of other more specialized techniques that are often used to map quantitative traits. One motivation for these techniques is the difficulty of calculating the likelihood for a general pedigree. There is some evidence that one of these methods, the multipoint variance components method, is comparable in power to multipoint parametric linkage yet it is much easier to test for linkage using this method. Another attractive feature of these methods is that one needn't specify a genetic model.

A feature of methods that seek QTLs that differs from the linkage methods that have been discussed thus far is that they frequently look for a QTL in an interval of the genome (e.g. a region that lies between 2 markers that are separated by 10 cM). Such procedures are referred to as *interval mapping*. The rationale behind such an approach is based on the theory of polygenic variation. Traits that exhibit quantitative variation among individuals have been shown in some animals to be attributable

Table 6.2 *Genotypes and quantitative phenotypes*

genotype	mean of phenotype
$A_1 A_1$	$\mu + \alpha$
$A_1 A_2$	$\mu + \delta$
$A_2 A_2$	$\mu - \alpha$

to the action of a major gene and many other, minor genes (or polygenes). Moreover, these polygenes tend to exist in clusters on the genome. This has been demonstrated in *D. melangaster* for the number of bristles on the organism. If a locus has only a small impact on altering the probability of having some phenotype, one will have difficulty detecting this locus. However, if one has a method that assesses alterations due to differences in larger genomic regions, one may be able to identify these clusters of polygenes. An additional complication that arises from considerations of QTLs that is remedied via interval mapping is that a QTL will reside a certain distance from a marker and have an effect on the trait with some quantitative magnitude. These 2 effects are necessarily confounded in a single locus approach. However, by considering intervals that lie between markers, the strength of a linkage signal can be interpreted as a genetic effect due to the QTL. Finally, these methods have been shown to have greater power than methods that just consider single loci (Goldgar, 1990).

6.5.1 Haseman Elston regression

The Haseman Elston regression method (Haseman and Elston, 1972) works with sib-pair data. For each sib-pair, we determine the proportion of alleles IBD at a given locus. Then we regress the squared difference between the sibling's trait values on the proportion of alleles shared IBD. Alternatively, we may regress this squared difference on estimates of the proportion of alleles IBD averaging over a number of markers, or over a chromosomal segment. The basic idea is that for a marker near a QTL, sibs with similar phenotypes should share many alleles IBD. Since all sibships are assumed independent, we can use the usual theory of least squares (see Section 2.3) to obtain parameter estimates and conduct inference. Interpretation of these results requires a model for the effect of the QTL on the trait, hence we briefly discuss these models next.

When discussing quantitative traits, it is often useful to differentiate between additive genetic effects and dominant genetic effects. Consider a QTL with 2 alleles, A_1 and A_2. We suppose that the mean value of the quantitative trait given the genotype at the QTL is as given in Table 6.2. We call δ the *dominance effect* and α is the *additive effect*. These effects are random effects, i.e. they represent realizations of a random variable. For example, if we suppose that there is no dominance effect (so that $\delta = 0$) and let y_{ij} represent the trait value for subject i in family j, the model for such effects can be represented as

$$y_{ij} = \mu + \alpha_j + \epsilon_{ij}$$

where

$$\alpha_j \sim N(0, \sigma_\alpha^2)$$

and ϵ_{ij} are iid $N(0, \sigma^2)$ errors. Notice that there will be no genetic effect if $\sigma_\alpha^2 = 0$ since in that case $\alpha_i = \alpha_j$ for all i and j (since there is no variance) and consequently $\alpha_i = 0$ for all i (since the mean of the set of α is zero). Finally, note that we have assumed a family specific random effect, hence this allows for heterogeneity between families regarding the size of the genetic effect. If we assume there is no dominance effect then it can be shown that the expected value of the slope in the regression equation is $-2(1-2\theta)^2 \sigma_\alpha^2$, where θ is the recombination fraction between the marker and the QTL. Hence if we reject the hypothesis that the slope is zero we conclude that $\theta \neq \frac{1}{2}$ and $\sigma_\alpha^2 \neq 0$, i.e. the marker is near a QTL which impacts the phenotype. Note that a confidence interval for the regression slope can not be used to estimate the recombination fraction or the additive genetic variance. Despite this, there are methods (first presented by Haseman and Elston, 1972) that allow one to estimate each of these parameters if one needs estimates (for genetic mapping for example).

6.5.2 Variance components models

Variance components models use the correlation for some trait as it depends on familial relatedness to map genes. These flexible methods were developed in Almasy and Blangero (1998) and refined and applied in Iturria and Blangero (2000), Blangero, Williams and Almasy (2000) and Comuzzie et al., (2001) (see also early work from Fulker, Cherny and Cardon, 1995). We will suppose that we are considering the case of interval mapping, although one can consider if a marker is linked to a disease gene using this approach. If we let y_{ij} denote the outcome variable for subject i in family j, then we suppose

$$y_{ij} = \mu + \alpha_j + \alpha_j^* + \epsilon_{ij}, \tag{6.1}$$

where α_j represents the additive genetic contribution for family j from the interval being examined and α_j^* represents the contribution from QTLs outside this interval. We assume α_j, α_j^* and ϵ_{ij} are uncorrelated random variables with means of zero and variances σ_α^2, $\sigma_{\alpha^*}^2$ and σ_ϵ^2 respectively. Since these 3 variables are uncorrelated the phenotypic variance is the sum of the variance of these 3 random terms. The sum of the variances of α and α^*, $\sigma_\alpha^2 + \sigma_{\alpha^*}^2$ is called the *total genetic variance* and the ratio of the total genetic variance to the total variance, $\sigma_\alpha^2 + \sigma_{\alpha^*}^2 + \sigma_\epsilon^2$, is called *heritability*. The size of the heritability is an indication of the extent to which a trait has genetic determinants.

We estimate these variance components by examing the phenotypic covariance of related individuals. Based on the linear model in equation 6.1 we can express the phenotypic covariance between 2 individuals in the same family as

$$\text{Cov}(y_{ij}, y_{i'j}) = R_{ii'}(j)\sigma_\alpha^2 + 2\Theta_{ii'}(j)\sigma_{\alpha^*}^2, \tag{6.2}$$

where $R_{ii'}(j)$ is the proportion of the chromosomal region under investigation shared IBD between individuals i and i' in family j and $2\Theta_{ii'}(j)$ is the coefficient of coancestry between these same 2 individuals. Below we discuss these 2 terms further. Given the set of $R_{ii'}(j)$s and $\Theta_{ii'}(j)$ for $j = 1, \ldots, N$, if we suppose there are n_j subjects in family j, and let y_j denote the vector of measurements $y_{1j}, \ldots, y_{n_j j}$, then we assume y_j is multivariate normally distributed with mean vector μ_j and covariance matrix (with elements given by equation 6.2 which depends on σ_α^2) Σ_j, hence we can express the likelihood of the data as a function of the variance parameter σ_α^2. Here, if we let $\mu = (\mu_1, \ldots, \mu_N)$, the likelihood is given by the expression

$$L(y|\mu, \sigma_\alpha^2, \sigma_{\alpha^*}^2, \sigma_\epsilon^2) = \prod_{j=1}^{N} (2\pi)^{-n_j/2} |\Sigma_j|^{-\frac{1}{2}} \exp\left\{-\frac{1}{2}(y_j - \mu_j)' \Sigma_j^{-1}(y_j - \mu_j)\right\}.$$

Then we can use the usual maximized likelihood ratio test to test the hypothesis that $\sigma_\alpha^2 = 0$. If we reject this hypothesis then we conclude that there is a QTL in the chromosomal region of interest. These models can be extended to include dominance effects and effects of covariates. For example, if we let $\mu_j = X_j \beta$ for a matrix X_j that holds covariate information for family j and a vector of regression coefficients, β, then we can allow covariates to impact the outcome variable. This would allow us to test for a genetic effect on the outcome while controlling for differences in environmental factors.

Coancestry

The coefficient of coancestry is a measure of the relatedness in terms of how close 2 subjects are in a pedigree. If 2 subjects are related one can count the minimal number of steps, m, one must take in a pedigree so that the 2 subjects have a common ancestor. In this case (provided there is only one path between the subjects) the coefficient of coancestry is $\frac{1}{2^{m+1}}$ and we expect that 2 such subjects would share a proportion of $2\frac{1}{2^{m+1}}$ of their genomes on average. For example, a parent and offspring are separated by one step in a pedigree, so that $m = 1$ and we find that the coefficient of coancestry is just $\frac{1}{2^{m+1}} = \frac{1}{4}$ and that these 2 subjects share $\frac{1}{2}$ of their genetic material. More formally, the coefficient of coancestry Θ_{ij}, is the probability that an allele at a locus selected randomly from individual i and an allele randomly selected from the same autosomal locus from individual j are IBD. When $i = j$ and we ignore the possibility of inbreeding then $\Theta_{ij} = \frac{1}{2}$. This probability depends solely on how related 2 individuals are: for example if i and j are sibs then $\Theta_{ij} = 1/8$. One can compute Θ_{ij} for all subjects in a pedigree recursively. First, one numbers all subjects so that all parents precede their children. The matrix of Θ_{ij} is constructed starting with the upper left corner. Supposing that we can ignore inbreeding, then $\Theta_{11} = \frac{1}{2}$ since this person is a founder. Then for $i > 1$ set $\Theta_{ii} = \frac{1}{2}$ if the person has no parents in the pedigree and set $\Theta_{ij} = 0$ for $j < i$ (due to the numbering convention). If i does have parents in the pedigree and they are numbered k and l, then set $\Theta_{ii} = \frac{1}{2} + \frac{1}{2}\Theta_{kl}$ since we are equally likely to select the same allele 2 times or the allele coming from the maternal and paternal source once. Then one fills in the other elements of the partial row and column for i by using the recursion

$$\Theta_{ij} = \Theta_{ji} = \frac{1}{2}(\Theta_{jk} + \Theta_{jl}).$$

6.5.3 Estimating IBD sharing in a chromosomal region

There are a number of methods for estimating R_{ij}, an unobservable random variable in the context of interval mapping. Here we will describe the method devised by Guo (1994). First, consider the extent of genetic sharing that occurs between 2 offspring in some genomic region (i.e. a part of a chromosome) on the chromosome that both siblings received from the maternal source, denoted ρ_M. This is unknown but we can estimate it if we have information on a set of markers that cover the genomic region. In addition there is a proportion that is shared on the chromosome that came from the paternal source, ρ_P. The total extent of genomic sharing between the siblings is then given by $\frac{1}{2}(\rho_M + \rho_P)$. The usual practice for dealing with this unknown is to obtain a point estimate of ρ_M and ρ_P, average these, then use this average as if it were known in all expressions that involve the proportion of genetic material that is shared IBD.

To estimate the extent of genomic sharing IBD in some region over which we have multiple markers, we first break the interval up into subintervals so that these subintervals are all flanked by markers and themselves contain no markers. Note that a complication arises if we are interested in estimating the extent of sharing between a marker that is the closest marker to the telomere (or centromere) and the telomere (or centromere), but this can be accounted for. If we suppose that we can assess if the siblings are IBD at each marker locus, then to determine the extent of sharing over the entire region, we need only determine what is the proportion of genomic sharing in each of the subintervals that are flanked by our informative loci. We then add over all subintervals to determine the extent of genomic sharing over the entire interval. To this end, we assume that there is no interference and that the process that led to the formation of the gametes that gave rise to the offspring is independent across offspring. Then, for each sibling, we suppose that there is an unobserved 2 state Markov chain, $h(t)$, that is in state 0 at location t if the genetic material at that location came from the maternal source and is 1 if it came from the paternal source (this is a continuous version of the inheritance vector). An illustration of such a function is provided in Figure 6.1. Since there is no interference, we can use Haldane's map function to give the following expression for the transition matrix

$$\frac{1}{2}\begin{pmatrix} 1 + e^{-2t} & 1 - e^{-2t} \\ 1 - e^{-2t} & 1 + e^{-2t} \end{pmatrix}$$

whose elements are referred to as $p_{ij}(t)$ for $i = 1, 2$ and $j = 1, 2$. If we define $\delta(u, v) = 1$ when $u = v$ and let $\delta(u, v) = 0$ otherwise and use $h_1(t)$ and $h_2(t)$ to represent the Markov chains for 2 siblings, then an expression for the proportion of

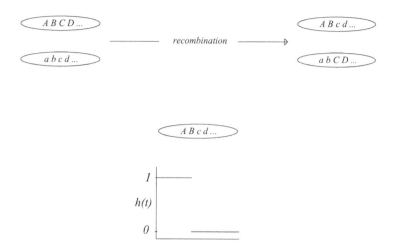

Figure 6.1 *An example showing a recombination. The 2 ellipses on the upper left represent 2 copies of a chromosome (the top came from this person's maternal source) while the 2 ellipses on the upper right correspond to the 2 copies of this chromosome after recombination. Letters correspond to loci and the different alleles are distinguished by case. The ellipse in the middle represents the copy of this chromosome that was transmitted by the parent to an offspring and the plot on the bottom shows the h-function for an offspring that receives this chromosome.*

a chromosome IBD over the interval $[0, C]$ is given by

$$R(C) = \frac{1}{C} \int_0^C \delta\big(h_1(t), h_2(t)\big)\, dt.$$

Now, suppose we have marker information at the ends of this interval so that we know $h_1(0), h_2(0), h_1(C),$ and $h_2(C)$. Then we estimate the proportion of the chromosome that is IBD with the expected value of this proportion, which is given by

$$E\big[R(C) \mid h_1(0), h_2(0), h_1(C), h_2(C)\big],$$

which we will denote as $E\big[R(C) \mid m\big]$ where m stands for the marker data at the ends of the interval. For example, suppose that $h_1(0) = 0, h_2(0) = 0, h_1(C) = 0$ and $h_2(C) = 0$. Then we have

$$E\big[R(C) \mid m\big] = \frac{1}{C} \int_0^C E\Big[\delta\big(h_1(t), h_2(t)\big) \,\Big|\, m\Big]\, dt,$$

but noting that $E[\delta] = P(\delta = 1)$ along with

$$P\Big(\delta\big(h_1(t), h_2(t)\big) = 1\Big) = \sum_{k=0}^{1} P\big(h_1(t) = k\big)P\big(h_2(t) = k\big),$$

(which follows from the fact that the processes that created each of the gametes that gave rise to the offspring are independent) we find that

$$E\big[R(C) \mid m\big] = \frac{1}{C} \int_0^C \sum_{k=0}^{1} P\big(h_1(t) = k \mid m\big)P\big(h_2(t) = k \mid m\big)\, dt.$$

Next, an application of Bayes theorem yeilds the following expression for the factors inside the sum

$$P(h_i(t) = k|h_i(0) = 0, h_i(C) = 0) \quad = \quad \frac{P(h_i(C) = 0|h_i(t) = k, h_i(0) = 0)}{P(h_i(C) = 0|h_i(0) = 0)}$$
$$\times \quad P(h_i(t) = k|h_i(0) = 0),$$

so if we then use the Markov property we find that

$$P(h_i(t) = k|h_i(0) = 0, h_i(C) = 0) \quad = \quad \frac{P(h_i(C) = 0|h_i(t) = k)}{P(h_i(C) = 0|h_i(0) = 0)}$$
$$\times \quad P(h_i(t) = k|h_i(0) = 0).$$

Now note that the factors on the right hand side are given by the elements of the transition matrix for the Markov chain that indicates the parental source of the genetic material in the chromosome, namely $P(h_i(C) = 0|h_i(t) = k) = p_{k0}(C - t)$, $P(h_i(t) = k|h_i(0) = 0) = p_{0k}(t)$ and $P(h_i(C) = 0|h_i(0) = 0) = p_{00}(C)$. This implies that

$$E\big[R(C) \mid m\big] = \frac{1}{C} \int_0^C \sum_{k=0}^{1} \frac{\big[p_{0k}(t)p_{k0}(C - t)\big]^2}{\big[p_{00}(C)\big]^2}\, dt.$$

If we then substitute the values from the tranistion matrix into this expression using Haldane's map and let $\theta = \frac{1}{2}(1 - e^{-2C})$ (i.e. the recombination fraction beween the markers that flank the region) we find (after some algebra)

$$E\big[R(C) \mid m\big] = \frac{1}{2} + \frac{\theta}{4C(1 - \theta)} + \frac{1 - 2\theta}{4(1 - \theta)^2}.$$

Other cases (i.e. configurations of the alleles at the flanking markers) can be dealt with similarly and the results for all 16 possible cases (including missing data at one of the markers) have been determined by Guo (1994).

6.6 A case study: dealing with heterogeneity in QTL mapping

While we have discussed tests for heterogeneity and linkage in the presence of heterogeneity, many approaches have been suggested for dealing with heterogeneity. Here we provide a detailed example that investigates one of these approaches. Here we consider the problem of identifying susceptibility loci for a complex disease that may exhibit locus heterogeneity for at least one of the genes involved in the disorder (see Ekstrøm and Dalgaard, 2003) for more on heterogeneity in the context of quantitative traits). This method was first proposed and applied in Reilly et al. (2007). We focus on quantitative traits as this provides a natural framework for studying complex diseases. The method was developed in the context of seeking genes that increase susceptibility to asthma, hence we also discuss the results from applying the method to this widespread condition. It is widely accepted that asthma and its associated quantitative traits are genetically complex (Daniels et al., 1996) and genetically heterogeneous (Hoffjan, Nicolae and Ober, 2003). The goal here is to identify a major gene involved in a quantitative phenotype that is modified by other genes and environmental factors that are heterogeneous. We would expect that there would be some familial environmental sources and polygenes that vary from family to family, but we would want to localize the common major genes since identification of such genes would allow detection and/or treatment of most with the disorder. Rather than develop methods for analyzing genotypic data, we here address the problem by attempting to define quantitative phenotypes based on a collection of variables that are measured for each subject. The constructed phenotypes will then be subjected to a genome scan using variance components linkage methods.

To accomplish these goals, we investigate the use of family information, although not genotypic information, for the construction of phenotypes. Below we present a model that allows incorporation of this information. The basic idea behind the model is best illustrated with an idealized example. Consider Figure 6.2. This figure represents data for affecteds (represented with darkened symbols) and unaffecteds (represented with undarkened symbols) for 3 families (the different symbols indicate family membership) for 2 continuous variables (the 2 coordinate axes in the figure). Here we suppose that there is a single major gene that is consistent across families, but there are 2 varieties of polygenes that modify the action of this gene (families 1 and 2 have the same modifying genes while family 3 has a distinct set). No linear transformation of these variables will result in a variable which will allow separation of the affecteds and unaffecteds, but if we measure the distance of each individual to the center of the cluster of the affecteds in his or her family, we obtain a variable which has the property that small values are associated with the disease allele for the major gene.

Our data were originally collected as part of the Collaborative Study on the Genetics of Asthma (CGSA, 1997). The present study used only the 27 multigenerational Caucasian families that were collected in Minnesota. These families had 169 asthmatic members (as defined below), 347 who were not asthmatic and 129 for whom the diagnosis was unavailable. Pulmonary function data (which quantifies lung health) were available on 619 individuals, but only the 456 phenotyped subjects who also had genotype data were used in the linkage analysis (there were not genotypic data available for all 619 subjects due to decisions regarding which matings were potentially informative for the phenotype used in the original study).

For the CSGA, families were ascertained through two asthmatic siblings. The fami-

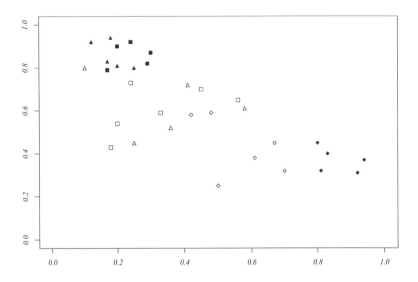

Figure 6.2 *An idealized example of a plot of 2 quantitative variables that are under the control of a major gene common to all affecteds but modified by factors that vary with family. In the figure the 3 different shapes correspond to 3 different families and a filled in shape indicates an affected.*

lies were "expanded" to permit recruitment of other relatives either by (a) extending the families through asthmatic relatives or (b) including no more than one unaffected relative to permit a lineage to incorporate other relatives with asthma. The quantitative data available arose from experiments designed to measure lung and airway health. Our basic data for the construction of phenotypes here was to use the logarithm of the percent predicted of the following variables: volume exhaled during the first second of a forced expiratory maneuver (FEV1), forced expiratory vital capacity (FVC), maximum expiratory flow when half of the FVC has been exhaled (FEFM) and forced expiratory flow rate over the middle half of FVC (FF25). While other transformations were considered, the logarithm led to the greatest amount of symmetry in the marginal distributions. Percent predicted refers to an observed value as a percentage of the predicted value given height, sex, and age; thus our analysis controls for sets of confounders that have an established influence on lung function measurements.

Here we describe a probability model for the trait values corresponding to the mechanism just described. The model uses a 2 level mixture structure. Suppose there are K possible genetic sources of some disease and we have N families with data on individuals. A genetic source is a high-risk genotype involving one or more loci. Furthermore, assume an affected individual has only one of the genetic sources, and all affected individuals in a family have the same genetic source. Our model supposes that the trait values for an individual arise from a family specific mixture distribution with 2 components: a diseased component and a healthy component. Supposing the distribution for those with the disease in family j has mean μ_j, we further suppose that these μ_j arise from another mixture model, reflecting the different genetic sources of the disorder. Note that we assume the number of genetic sources is less than the number of families. If one suspects that each family has a distinct genetic source, one could just conduct a distinct linkage analysis for each family. In some situations, we may have further data on the subjects than just the trait values used for constructing the quantitative phenotype, as in the asthma data set investigated here, hence we allow the mixing proportions in the family specific mixture model to be subject specific, depending on a set of other variables.

In general, if y_{ij} represents the p-vector of quantitative variables for subject i in family j, x_{ij} is a set of covariates for this subject indicating if the subject is likely affected, β is a vector of regression coefficients giving the effect of each element of x_{ij}, $f_j(\mu_j)$ and $h_k(\eta_k)$ are both densities on a Euclidean space of dimension p with means μ_j and η_k respectively, g_j is another density on the same space, λ_{ij} is an indicator variable which is one when subject i in family j is asthmatic, and weights ϕ_k satisfying $\sum_{k=1}^{K} \phi_k = 1$, then we can write the above model in the following form:

$$y_{ij} \sim \lambda_{ij} f_j(\mu_j) + (1 - \lambda_{ij}) g_j$$
$$\mu_j \sim \sum_{k=1}^{K} \phi_k h_k(\eta_k)$$
$$\lambda_{ij} \sim \text{Ber}(x_{ij}'\beta).$$

The above model describes the marginal distribution for a single individual. A full likelihood based treatment (which is not pursued here) needs to also take account of correlation in the y_{ij} that is induced by familial relationships. This would be accomplished by using the kinship information for related subjects (unrelated subjects are modeled as independent). Given this framework, we define our quantitative variable as

$$z_{ij} = \| y_{ij} - \eta_k \|_2^2,$$

where k is the index of the mixture component to which family j belongs (and $\| x \|_2$ is the Euclidean norm of the vector x). We use this as our variable for a genome screen because we want to measure how far an individual is from the typical affected in families that have the same genetic source as the family to which this individual belongs.

Note that the phenotypes z_{ij} depend on a set of parameters (namely η_k for $k = 1, \ldots, K$ and cluster membership for all subjects), hence we must estimate these parameters to determine the phenotypic values. A full likelihood based approach would involve many assumptions about the densities in the previous section, and even if one would provisionally assume normality, issues of identifiability would likely arise given the full generality of the parameterization there (for example, would each family have its own covariance matrix describing the covariance of y_{ij} for a common j). Even then computation is daunting (due to the 2 level mixture model combined with estimation of β). Thus we propose an algorithm that is based on the model which combines common sense estimators with heuristic clustering algorithms.

Here we suppose information is available about likely cases (but this information alone is not sufficient to detect linkage). First, we use this information about likely cases to identify "affecteds." Then we compute the mean of all p variables used for phenotype definition across all "affecteds" in a family. The resulting set of mean values (one set of p means for each family) can be thought of as characterizing the typical affected for each family (the typical family level affected). Denote these averages $\overline{x}_j$ for $j = 1, \ldots, N$, where $\overline{x}_j$ is a p-vector. Next we conduct a cluster analysis on these typical family level affecteds, $\overline{x}_j$. Cluster analysis is a set of techniques whose goal is to sort units into groups so that units within groups are similar in terms of the data we have on the units (see Chapter 14 for further details). If this suggests the presence of clusters, then, for each subject in the data set, we determine family membership and measure the distance of the p vector of data for that subject, y_{ij}, to the average of the $\overline{x}_j$'s in the cluster containing that subject's family. That is

$$z_{ij} = \| y_{ij} - \tilde{x}_k \|_2^2$$

where k is the index of the cluster to which the affected in family j belongs, and $\tilde{x}_k$ is the average of all $\overline{x}_j$ in that cluster. This provides us with a quantitative score for each subject in the study. This algorithm can be interpreted as an approximate solution to a special case of the general model. We used k-means clustering with increasing numbers of clusters. As the number of clusters is increased some of the clusters appear to be similar in terms of the means for objects in distinct clusters, hence fewer clusters are adequate. Examination of the results from the cluster analysis suggested the presence of 4 clusters.

To ensure that familial relationships had been specified correctly, we ran all pedigree data through the Graphical Representation of Relationships (GRR) program (Abecasis et al., 2001). This program allowed rapid confirmation of most relationships and easy detection of DNA sample mixups between or within families. We detected no errors. We then checked for Mendelian inconsistencies using PEDCHECK (O'Connell and Weeks, 1998) and we corrected errors by setting appropriate genotypes to missing. Finally, we ran MERLIN (Abecasis et al., 2002) with the "–error" option to identify further errors that appeared only as low-probability double recombinants and we again corrected likely errors by setting appropriate genotypes to missing.

Nonparametric variance components linkage analyses were conducted in SOLAR (Almasy and Blangero, 1998) but we used multipoint IBD probabilities computed in MERLIN and LOKI (Heath, 1997) to avoid the problems in SOLAR's implementation of the Fulker algorithm (Fulker, Cherny and Cardon, 1995). We used marker allele frequencies computed by MERLIN from founders in our pedigree data. When families were small enough (fewer than 27 bits) we used MERLIN's implementation of the Lander Green algorithm to compute multipoint IBD probabilities. For larger families, we used the Markov chain Monte Carlo (MCMC) algorithms implemented

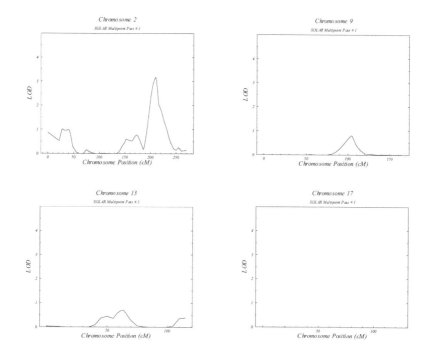

Figure 6.3 *LOD scores for all markers on 4 selected chromosomes included in present study.*

in LOKI with 1,000,000 iterations and a 50-50 mix of L and M steps. The results of the SOLAR genome scan for the derived quantitative phenotype for all chromosomes are shown in Figure 6.3. Initially, a peak LOD score of 3.17 was observed on chromosome 2 at marker D2S2944 which was located at 210 cM on our map. The peak was approximately 15 cM wide and was flanked by marker D2S1384 at 200 cM and marker D2S434 at 215 cM. The addition of four microsatellite markers in that region (D2S1782, D2S1369, D2S1345 and D2S1371) increased the peak LOD score to 3.40 at the same location. Figure 6.4 shows the LOD scores after the addition of these markers. We also conducted full genome scans in SOLAR for all four of the variables used to construct the derived phenotype. None of the four showed evidence for linkage anywhere in the genome. The highest LOD score among all four across all chromosomes was 0.72, and no trait had a substantial peak on chromosome 2. Thus the proposed method used a nonlinear mapping from a 4-vector to a scalar where there was no linkage signal from any component of the 4-vector and yet there was for the scalar.

The 2q33 region has been identified in other genome scans (The Collaborative Study on the Genetics of Asthma, 1997; Lee et al., 2002), but the peak at 210 cM has not been specifically identified in previous studies. Most other studies concentrate on the region of 2q33 where CD28 and CTLA4 are located, which is at approximately 200

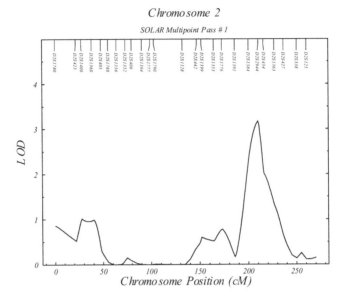

Figure 6.4 *LOD scores for chromosome 2 with the addition of 4 markers.*

cM. An investigation of known genes in this region reveals several promising candidates, some of which have shown evidence for linkage to phenotypes associated with asthma: inducible T-cell co-stimulator, cytotoxic T-lymphocyte-associated protein 4 (CTLA4), CD28 antigen and the interleukin 8 receptors alpha and beta.

6.7 Linkage disequilibrium

It is reasonable to suppose that some genetic disorders arise due to a mutation at some gene. A mutation can lead to an organism that doesn't produce the protein necessary for some biological function. Consider some individual who has undergone a mutation that leads to a dominant disease. If we use D to represent the mutant disease allele, and single digit numbers to represent marker alleles, then his genotype on say chromosome 4 is something like

chromosome 4, copy 1

$$1, 4, 7, 3, 8, D, 3, 7, 3, 6, 2$$

chromosome 4, copy 2

$$4, 2, 6, 2, 5, d, 2, 6, 1, 7, 8.$$

He will transmit one of these chromosomes to each of his offspring. Consider an offspring who is affected. Such an offspring must have received copy 1, but perhaps there was a crossover, so this individual's chromosome 4 from the affected parent would be something like

$$4, 2, 6, 3, 8, D, 3, 7, 3, 6, 2.$$

As generations come and go, one of the chromosomes in all of the affected individuals must have the disease allele. Those with the disease allele will also have the markers near the disease loci that the original family member with the disease had unless there is some recombination that results in another marker taking its place. That is, we would expect to find alleles 8 and 3 at the marker loci next to the disease locus in affected offspring of this individual, a state we refer to as *linkage disequilibrium*.

We define the *disequilibrium parameter*, δ, for the marker with allele value 8 in the above example as

$$\delta = P(D, 8) - P(D)P(8).$$

This parameter is not so useful as a measure of linkage disequilibrium because its minimal and maximal value depend on the allele frequencies. If there is no linkage disequilibrium

$$P(D, 8) = P(D)P(8),$$

and so we expect δ to be near zero (otherwise we expect it to be positive). Over time, the disequilibrium would decay at a rate depending on the recombination fraction between the disease and marker loci. If δ_t is the linkage disequilibrium at time t, then

$$\delta_t = (1 - \theta)^t \delta_0,$$

hence if one has an estimate of the recombination fraction and $P(8)$, one can estimate the time of the original mutation.

The disequilibrium parameter can be estimated by using sample proportions. This also indicates how to construct confidence intervals for δ. To test the hypothesis

$$H_0 : \delta = 0$$

one tests if the disease haplotype (i.e. a set of markers associated with the disease phenotype) occurs in significantly different proportions in the affected versus the non-affected populations. If haplotypes are not so easily constructed, one must average over phase. These tests fall into the category of association tests since they look for an association between haplotypes and disease phenotypes where the haplotype is common across all affecteds in a population.

Linkage disequilibrium can arise for reasons other than a linkage of a marker to a loci where a mutation took place. For example, if you use a set of unrelated individuals it may appear that there is linkage disequilibrium because your sample is drawn from a heterogeneous populations with sub-populations that differ in terms of gene frequencies and disease frequency. Even with information on ethnicity, the existence of subpopulations is problematic because the subpopulations can be quite subtle.

Given this, linkage disequilibrium mapping will be most useful for mapping when the population under study is a small isolated group with extensive intermarrying. In particular, this method can be problematic for analyzing heterogeneous populations with individuals coming from a variety of sources. When conducting association studies with heterogeneous populations one must use a test statistic that tests for association and linkage at the same time or adjusts the association test for population stratification.

6.8 Association analysis

Association analysis is a term used to refer to methods that aim to establish if there is an association between a trait and the occurrence of a certain allele at some marker in a population of individuals. This would arise if the marker were very close to a trait locus (so that only a large number of crossovers would separate the 2 loci) and the trait arose as a mutation in a single individual far in the past. Depending on the nature of the phenotypic data it is usually straightforward to test for an association between a phenotype and the occurrence of a particular allele. For example if the phenotype is dichotomous (e.g. healthy or diseased), one can construct a contingency table and test for an association between the occurrence of a particular allele and a trait using Pearson's χ^2-test. However, we will see that more sophisticated approaches are necessary when examining heterogeneous populations.

The primary difference between association analysis and linkage analysis is that the former will only find associations between the occurrence of an allele and a trait if the marker locus is very close to the trait locus. On the other hand, linkage analysis will find an association between the occurrence of a marker allele and the trait locus over a long range. This property makes the 2 approaches complementary since one can use linkage analysis to determine roughly where the trait locus is, then use association analysis in this region to conduct what is known as fine mapping. While this approach is still popular, more recent research has focused on using very high density sets of SNPs (on the order of 1 million) and testing for an association between SNP genotypes and disease traits (this approach is referred to as a genome wide association analysis (GWAS)). This raises the problem of multiple testing, about which we say more in Chapter 13. Also note that in linkage analysis the marker allele that segregates through a family with a disease allele can vary across families whereas in association analysis we require the same marker allele to be in linkage disequilibrium with the disease allele for all affecteds.

6.8.1 Use of family based controls

As mentioned, one of the major problems with tests of linkage disequilibrium that use ordinary statistical tools on unrelated subjects is the unaffecteds are unrelated, hence the haplotypes of these individuals make for poor "control" haplotypes. A better way to define a control population would be to use unaffected individuals from the same family, although this isn't always feasible. For now we will suppose we have trios of parents and an affected child. Consider the problem of whether some allele, denoted A_1, at a marker is in linkage disequilibrium with with some disease. For each family, define the disease genotype (or haplotype if we consider more than 1 marker) to be the genotype at the marker in question for the affected child. This disease genotype consists of 2 alleles. The control genotype consists of the 2 alleles not transmitted to the affected child. For example, if the parental genotypes are $(A_1|A_2)$ and $(A_3|A_4)$ and the child has genotype $(A_1|A_4)$, then the disease genotype for this family is $(A_1|A_4)$ and the control genotype is $(A_2|A_3)$.

Haplotype relative risk

Comparing cases and controls based on a study design that samples by affectation status is called a retrospective study. A standard measure of the risk associated with some risk factor (such as having a particular allele at some marker locus) is the relative risk. The relative risk is the ratio of the probability of being a case given that one has some risk factor to the probability of being a case given that one doesn't have the risk factor. To estimate

$$P(\text{case}|\text{risk factor})$$

and

$$P(\text{case}|\text{don't have the risk factor})$$

it would appear that the best strategy is to obtain a group of subjects that are positive for the risk factor and a group of subjects that don't have the risk factor then follow them to determine what proportion become a case. The drawback of such a prospective study is that if the disease is rare one needs to follow many subjects in order for there to be enough cases. Moreover it is easier to sample subjects by disease status since affected subjects interact with doctors frequently and are thereby easier to recruit into a study. The odds ratio is another measure of the risk associated with exposure to a risk factor. With the odds ratio, we replace the probability of each of the 2 events with the odds on that event (recall that if p is the probability of some event, the odds on that event are defined to be $p/(1-p)$). For rare diseases the odds ratio and the relative risk are approximately equal. Moreover, odds ratios can be estimated from retrospective study designs, hence we can obtain an approximation to the relative risk due to some exposure from a retrospective study design by using the odds ratio. Further information regarding prospective and retrospective designs and their analysis can be found in Fleiss (1986).

As a direct extension of the usual relative risk in the context of a case control study, the *haplotype relative risk* associated with allele A_1 is defined as

$$\text{HRR} = \frac{P(\text{disease genotype}|A_1)}{P(\text{disease genotype}|\text{not } A_1)}.$$

The test for linkage of the allele A_1 to the disease locus in this context is

$$H_0 : \text{HRR} = 1.$$

Table 6.3 *Relationship between occurrence of an allele and a dichotomous trait.*

	has allele A_1	doesn't have allele A_1	total
case genotype	w	x	n
control genotype	y	z	n
total	$w + y$	$x + z$	$2n$

We can test this hypothesis by setting up the table shown in Table 6.3 and conducting Pearson's χ^2 test for independence. This is because the alleles that are transmitted to the affected offspring and those that aren't transmitted are independent when either there is no disequilibrium (the hypothesis of interest) or when the recombination fraction is zero (as shown by Ott, 1989). But this latter fact also implies that if we reject the hypothesis that the disequilibrium is zero we cannot conduct valid inference for the HRR unless the recombination fraction is zero, i.e. the marker is right next to the disease locus. As is usual in retrospective study designs, the HRR is estimated assuming that the disease is sufficiently rare so that the odds ratio and the relative risk are equal, hence the estimate of HRR is given by the usual expression for an odds ratio $\frac{wz}{xy}$. However, as noted above, this is only a useful estimate when the recombination fraction is zero if we reject the hypothesis of no association.

Construction of the table is straightforward. If there are n families, then there are n affected children, hence there are n disease genotypes and n control genotypes (so there are $2n$ entries in the table). If there is linkage disequilibrium between the allele and the disease gene, we would expect the concordant cells of the table to fill up. Moreover, under the assumption of linkage equilibrium, alleles segregate independently of one another, so we can think of this table as representing data from $2n$ independent observations. Thus we can apply Pearson's χ^2 test to the table as suggested previously. For the example where the parental genotypes are $(A_1|A_2)$ and $(A_3|A_4)$ and the child has genotype $(A_1|A_4)$, we put an observation in the cell representing the A_1 is in the disease genotype, and an observation in the cell corresponding to a not A_1 in the control genotype. If there is linkage disequilibrium between allele A_1 and the disease gene, we would expect to see observations pile up in the cells in which allele A_1 is in the disease genotype. Note that if neither parent has allele A_1 the mating is completely non-informative about this allele, however this mating would contribute an entries to cells x and z. Similarly, a parent that is homozygous for A_1 will always contribute an entry to cell w. Such a parent is not doubly heterozygous and such parents can't be informative about linkage.

Haplotype-based haplotype relative risk

Another test based on the use of control genotypes is best understood by considering the events that lead to transmission of an allele for each patent. Consider Table 6.4 which shows the possible outcomes for transmitting alleles to offspring. For example,

Table 6.4 *Relationship between occurrence of an allele and a trait.*

	control A_1	control $\overline{A}_1$	total
case A_1	a	b	w
case $\overline{A}_1$	c	d	x
total	y	z	n

suppose a pair of parents have genotypes $(A_1|A_2)$ and $(A_3|A_4)$, while the affected offspring has genotype $(A_1|A_3)$. In this case, the parents have transmitted an A_1 and not transmitted a pair of not A_1, hence they provide an entry into cell b in Table 6.4. In this manner Table 6.4 gets filled up with a total of n entries.

The idea behind the *haplotype-based haplotype relative risk*, HHRR, is that parental genotypes are independent under the null hypothesis of no association between the allele in question and the trait under investigation. Since the parental genotypes are independent under the null hypothesis, we can treat each parent independently of the other and look at the fate of each of the 4 alleles at the marker locus in terms of being transmitted to the case offspring. For example, the parent with genotype $(A_1|A_2)$ has transmitted A_1 and not transmitted A_2 to the case genotype hence this parent will place an entry into cell b, while the other parent has transmitted A_3 to the case and not transmitted A_4 hence this parent adds a count to cell d in Table 6.4. Note that now the total number of entries in this table will be $2n$ and if we then convert back to a table of the form Table 6.3 we have a 2 by 2 table with $4n$ entries. If we then conduct Pearson's χ^2 test of no association on the resulting table we obtain the test statistic for the HHRR. This has been shown to be more powerful than the test associated with the HRR or the transmission disequilibrium test, which is discussed in the next section (Terwilliger and Ott, 1992).

The transmission disequilibrium test

Another test for testing the hypothesis of no linkage via linkage disequilibrium is known as the *transmission disequilibrium test*, TDT (Speilman, McGinnis and Ewens, 1993). This test is much like the HRR, but it ignores families where one parent is homozygous for the allele suspected to be in association with the disease gene (since these obscure our view of linkage disequilibrium) and matings where neither parent has the allele suspected of being in linkage disequilibrium with the trait locus. Recall both of these sorts of matings were used in the context of the HRR. We think of each parent as contributing either the allele in disequilibrium with the disease genotype (i.e. the case allele) or not (the control allele) and create a table like Table 6.4 except we don't treat parents independently as was done in the context of developing the HHRR (so there are a total of n entries in the table, one for each mating). Since the parents are matched, the appropriate test is to apply McNemar's χ^2 test for Table 6.4.

We expect the numbers in the off-diagonal of the table to fill up under the null so we compare these. When we work out the sample variance of this difference, we arrive at the test statistic, namely

$$\frac{(c-b)^2}{c+b}.$$

If we let p_D represent the probability of observing the disease allele then it can be shown that (see, for example, Liu, 1998)

$$\mathrm{E}(c-b) = 2n(1-2\theta)\delta/p_D.$$

Hence the expected value of the difference is zero if $\delta = 0$ or $\theta = \frac{1}{2}$, i.e. if there is linkage disequilibrium or if there is linkage (or if both are true). For this reason, it is recommended that linkage disequilibrium first be established using other methods (which is easy since we just have to detect association) then one tests for linkage using the TDT. In such a context, it has been shown that the TDT is a powerful test for establishing linkage (Guo, 1997).

There have been many extensions of the TDT. For example, if parental genotypes are missing, we may be able to at least partially deduce them from the offspring genotypes. We can then compute a TDT statistic, but must account for the fact that we have reconstructed the genotypes (this test statistic is called the reconstruction combined TDT or RC-TDT, see Knapp, 1999). There are also extensions to the case where the outcome is quantitative as in Rabinowitz (1997) and Allison (1997). The family based association test (FBAT) includes most of these extensions as special cases, at least approximately (Horvath, Xu, and Laird, 2001). The FBAT can be thought of as a 2 stage procedure in which one first computes some statistic that tests for association. In the second stage one treats the offspring genotype data as random and computes the null distribution of the test statistic conditional on certain parts of the data (such as the observed traits and the parental genotypes). For example, if $T_{ij} = 1$ when subject j in family i is affected and $T_{ij} = 0$ otherwise, and if X_{ij} counts the number of a times a particular allele occurs in the genotype of subject j in family i for some marker, then if we set

$$S_i = \sum_j X_{ij} T_{ij},$$

the statistic S_i counts the number of times the allele under investigation appears in the genotype of an affected offspring in family i. The second stage then computes the null distribution of this statistic. By modifying the definition of T_{ij} we can test for an association between a quantitative trait and a marker allele. There are general approaches that have been developed to determine the variance of S_i that allow one to develop test statistics (Horvath, Xu, and Laird, 2001). There is widely available software that allows one to use the method.

6.8.2 Correcting for stratification using unrelated individuals

A number of methods have been proposed that attempt to correct for population stratification when analyzing data from unrelated individuals. The simplest approach would be to test for an association between the occurrence of an allele and the trait of interest controlling for the effect of ethnicity. For example, suppose the phenotype under investigation is a quantitative trait, denoted y_i for subject i, the genotypic data is from a SNP and for each subject we have data on ethnicity, which we will assume is dichotomous and represent with x_{1i} for subject i. For SNPs, there are only 3 possible genotypes since all subjects are either heterozygous, homozygous for the minor allele or homozygous for the major allele. Hence we could create 2 indicator variables to code the genotypes, say $x_{2i} = 1$ if subject i is heterozygous and 0 otherwise, while $x_{3i} = 1$ if subject i is homozygous for the major allele (so a subject that is homozygous for the minor allele will have $x_{2i} = 0$ and $x_{3i} = 0$). Alternatively, we could let x_{2i} be 0 or 1 depending on the marker genotype (where 0 is the heterozygous genotype). Such coding of the covariate information implies there is no dominance effect in this model. If we use the 2 indicator variable approach we could then fit the linear regression model

$$y_i \sim N(\beta_0 + \beta_1 x_{1i} + \beta_2 x_{2i} + \beta_3 x_{3i}, \sigma^2).$$

This model supposes that the mean value of the trait depends on ethnicity, but given this difference in the means, the genotypes have the same impact on the trait of interest. A more complex model would allow an interaction between ethnicity and genotype, i.e.

$$y_i \sim N(\beta_0 + \beta_1 x_{1i} + \beta_2 x_{2i} + \beta_3 x_{3i} + \beta_4 x_{1i} x_{2i} + \beta_5 x_{1i} x_{3i}, \sigma^2).$$

In this model, not only is there an overall difference in the mean between the 2 ethnicities, but the effect of the genotype depends on the subject's ethnicity (i.e. there is heterogeneity). Most approaches to controlling for population stratification use the first model that ignores the interaction. This is somewhat inconsistent with the usual interpretation of the source of linkage disequilibrium. For, suppose that a mutation that influences some trait came about in Africa that was transmitted to the offspring of the carrier of the mutation. Further suppose that this mutation is never found in non-Africans and one has a sample of subjects from Africa and another place. In this case, the trait influenced by the mutation will have a different mean value in the 2 subpopulations, but there will only be an association between the trait and markers near the location of the mutation in those Africans that received the mutation.

While we may not have data on ethnicity for a set of subjects, even if we do, it is usually too course of a variable to account for subtle structures within a given population. For example, as one traverses Europe, there is continuous variation in allele frequencies that doesn't obey political boundaries (Price et al., 2006). For this reason, a number of approaches have been developed to account for this unmeasurable population substructure. The first method that was proposed, called genomic control (Devlin and Roeder, 1999), attempted to correct the sampling distribution of a test statistic that is used to test for differences between genotypes. The 2 other primary methods attempt to control for the effect of population substructures by estimating the subpopulations and determine membership of subjects to these subpopulations using data from a large collection of markers. These markers are invariably SNPs as these methods are used in the context of conducting large scale, genome wide association studies.

The first of these methods that was proposed is called structured association analysis (Pritchard, Stephens and Donnelly, 2000). This method uses data on SNPs randomly distributed throughout the genome to find groups of subjects that are similar in terms of their genotypes for these SNPs. The method uses the model based clustering method discussed in Section 14.2.3. This determines the number of subpopulations (provided it is less than an upper bound set by the user) and the probability that each subject belongs to each subpopulation. One then uses these probabilities as covariates in a linear regression model in the same manner as if one had information on the ethnicity of each subject. While this method has been shown to work well in practice, the computational time necessary to use the method can make it difficult to use in large scale studies involving many markers. An alternative method that uses a statistical technique called principal components to find the clusters of subjects has been shown to perform as well as the structured association method and can be used with very large numbers of markers (Price et al., 2006).

6.8.3 The HAPMAP project

In an attempt to understand the pattern of linkage disequilibrium across the genome and to obtain information regarding the locations of SNPs in a variety of populations, a number of researchers have started the HAPMAP project. The data arose from the Yoruba people of Nigeria (30 trios, i.e. 30 parents and 1 grown offspring), 45 unrelated individuals from the Tokoyo, Japan area, 45 people from the Beijing, China area and 30 trios from the United States. This project involves genotyping millions of SNPs in each subject to estimate SNP frequencies in the various populations and estimate the level of linkage disequilibrium across the genome. By accessing the website for the project, one can find out about which SNPs are in different genomic areas. This is very useful for finding SNPs near candidate genes in association tests. Another application of these results is the identification of what are known as *tag SNPs*. When there is a high degree of linkage disequilibrium, knowledge of the allele at one locus will be informative for predicting the alleles at nearby loci. Thus if one is planning a study to detect an association between a trait and a locus, if there is high level of linkage disequilibrium one need only genotype one or a few SNPs in that area since these SNPs will reflect the alleles at other loci. The SNP that is used as a surrogate for other SNPs in this fashion is referred to as a tag SNP.

6.9 Exercises

1. Suppose at some marker, the parental genotypes are (A_1, A_2) and (A_1, A_3) and the 2 offspring of these parents have genotypes (A_1, A_2) and (A_1, A_3). How many alleles do these offspring share IBD? How many alleles do they share IBS?

2. Suppose at some marker, the parental genotypes are (A_1, A_1) and (A_2, A_3) and the 2 offspring of these parents have genotypes (A_1, A_2) and (A_1, A_3). Show that one can't determine the value of the number of alleles shared IBD, however the expected value of this quantity is $\frac{1}{2}$.

3. Find estimates of the standard deviation for the mean and proportion tests. Use these estimates to construct test statistics that have standard normal distributions under the relevant null hypothesis.

4. If the phenocopy rate for a recessive trait is 0.3, then what are the IBD probabilities for a marker tightly linked to the disease gene for 2 affected sibs? (Assume the genotypes of the parents are (A_1, A_2) and (A_3, A_4), both parents are heterozygous at the disease locus.)

5. Find the coefficients of coancestry for the pedigree in Figure 4.1.

6. Derive the expression for the proportion of the chromosomal region that is IBD using the approach of Guo for the case presented in Section 6.5.3.

7. Derive the expression for the proportion of the chromosomal region that is IBD using the approach of Guo for the case when $h_1(0) = 0$, $h_1(C) = 1$, $h_2(0) = 1$ and $h_2(C) = 0$.

8. What is the maximal value of the disequilibrium parameter between 2 SNPs, and what major allele frequencies give this highest value?

9. Consider 2 SNPs in 2 distinct populations where the minor allele frequencies in one population are 0.3 and 0.2 and the corresponding minor allele frequencies in the other population are 0.15 and 0.1. If the 2 populations mix in equal proportions, what is the value of the disequilibrium parameter?

CHAPTER 7

Sequence Alignment

7.1 Sequence alignment

As mutation and selection drive evolution, DNA and genes change over time. Due to these forces, 2 species with a common ancestor will have similar, but non-identical genes in terms of base-pair sequence. This basic fact can be used for several purposes. For example, suppose you know the DNA sequence for a human gene but are unsure of the function of the gene. If you can find a gene with a similar sequence in a closely related species, then a reasonable conjecture is that the functions of the 2 genes are the same. Another application of this basic idea is in constructing phylogenetic trees. These are graphical representations that show how closely related a set of species are to one another. Such considerations often have medical applications: for example, by knowing how similar different strains of human immunodeficiency virus (HIV) are to one another we may have some idea about how effective a vaccine will be that has been designed using a certain strain. Yet another common application is in the recognition of common sequence patterns in DNA (the motif recognition problem). Recognizing common patterns in DNA or proteins can help determine where genes are in a sequence of DNA or proteins. We can also try to find locations where certain proteins bind to DNA if we suppose there is some sequence of DNA that is recognized by a protein.

These are some of the basic problems that drive the question: how similar are 2 bio-polymers. There are many other applications of these techniques. Previously we noted that while thinking of DNA as a linear molecule is often a useful abstraction, this is not a useful way to think about proteins. Hence comparison methods for proteins should consider not just the sequence of amino acids, but also the structure of the protein (see, for example, Levitt and Gerstein, 1998). Similarly, RNA can exhibit complicated 3 dimensional structures, although the structures are not as varied as protein structures.

Methods for comparing 2 or more biomolecules are typically referred to as methods of *sequence alignment*. The name comes from the basic idea that to see how similar 2 sequences are, we should write one down, then try to line up the other sequence with what we have written down. What we mean by "line up" is really the heart of the matter. Mutation of DNA or protein sequences is thought to occur through just 4 basic processes:

1. changes in a single nucleotide or amino acid

2. deletion of a single nucleotide or amino acid

3. addition of a single nucleotide or amino acid

4. copying of large segments from one location to another.

Differences we see in DNA across species are thought to arise through the combination of many such basic mutations and subsequent natural selection. Hence, if we are trying to line up sequences thought to have a common evolutionary origin, we should allow mismatches and gaps in the 2 sequences. Some mutations are selected against, for example, if a hydrophobic residue in a protein takes the place of a hydrophilic residue the protein may lose the ability to perform its function, hence we don't expect such mismatches when we line up the sequences. Other mutations may not affect the function of the biomolecule, and as such, may be consistent with a common progenitor. For these reasons, when we line up the sequences, we should consider some mismatches more severe than others, and we should also rate how severely gaps may lead to a non-functional biomolecule in terms of these mismatch penalties.

Sequences that share a certain degree of similarity are referred to as *homologous*, and we say the 2 sequences are *homologs* are homologs that arose due to gene duplication in an organism. After this gene duplication, one of the copies diverged from the common ancestor gene. *Xenologs* are homologs that resulted from horizontal gene transfer, i.e. a gene entered the genome of some organism from another organism. Orthologs tend to have common functions, whereas paralogs tend to have different functions. Xenologs may or may not have the same function. A gene is typically inferred to be a xenolog if the gene seems out of place in a genome. For example, the *GC content* (i.e. the percentage of the bases in a single stranded DNA molecule that are G or C) of a gene that is horizontally transferred into a species may depart strongly from the overall GC content of the genome into which the gene has been transferred.

7.2 Dot plots

The idea of the dot plot is quite simple: put 2 sequences along the margins of a table and enter a dot in the table if the 2 sequences have the same letter at that location. One can then graphically inspect the sequences to detect if there are long shared regions. Such regions will show up as a long diagonal string of dots in the plots. While this simple method would indicate stretches of shared sequence elements, it would not reflect the fact that some mutations affect the functionality of the resulting protein more than others.

A heuristic technique for comparing 2 sequences would be to count the longest diagonal stretch in the plot. One could then compare this value with what we expect if the agreement is just chance agreement. This simple idea is the basis for methods used to look for homology of a given sequence with all sequences in a database, as we discuss in Section 8.2.

As an example, consider the protein interleukin 7 (IL7) which is found in both the mouse and humans. IL7 is an important protein for the operation of the immune system. The amino acid sequence for these 2 homologous sequences can be found online (using the SWISS-PROT database), and the human version has the following amino acid sequence (the amino acid sequences are broken up into groups of 10 as is customary)

```
        10         20         30         40         50
MFHVSFRYIF GLPPLILVLL PVASSDCDIE GKDGKQYESV LMVSIDQLLD
        60         70         80         90        100
SMKEIGSNCL NNEFNFFKRH ICDANKEGMF LFRAARKLRQ FLKMNSTGDF
       110        120        130        140        150
DLHLLKVSEG TTILLNCTGQ VKGRKPAALG EAQPTKSLEE NKSLKEQKKL
       160        170
NDLCFLKRLL QEIKTCWNKI LMGTKEH
```

while the mouse version is as follows

```
        10         20         30         40         50
MFHVSFRYIF GIPPLILVLL PVTSSECHIK DKEGKAYESV LMISIDELDK
        60         70         80         90        100
MTGTDSNCPN NEPNFFRKHV CDDTKEAAFL NRAARKLQF LKMNISEEFN
       110        120        130        140        150
VHLLTVSQGT QTLVNCTSKE EKNVKEQKKN DACFLKRLLR EIKTCWNKIL
KGSI.
```

Note that the mouse version is slightly shorter than the human version and that both start with M (which is the amino acid for the start codon).

As Figure 7.1 shows, there is a long stretch of amino acids that is common to both versions of this protein for roughly the first 110 amino acids. After this, there appears to be an large insertion in the human version, but around the 150th residue in the human version the agreement between the 2 sequences resumes.

7.3 Finding the most likely alignment

In order to adopt a likelihood based method for aligning 2 biosequences, we must specify the likelihood for pairs of sequences, which we will denote y_1 and y_2, with

human

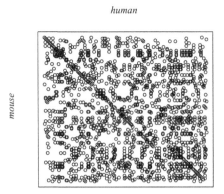

mouse

Figure 7.1 *An example of a dot plot. Here the human and mouse versions of the protein IL7 are examined for similarity.*

elements y_{ij} for $i = 1, 2$ and $j = 1, \ldots, n_i$. The set of possible values of y_{ij} is called the alphabet (for example, the alphabet for DNA is A, C, G, T). One simple approach is to suppose that the elements y_{ij} are iid observations from a multinomial distribution over the alphabet. While the iid assumption is not realistic in that we certainly don't think all possible sequences of amino acids of a fixed length are equally likely to be observed in living organisms (since such a sequence performs some function), and given evidence for serial dependence in sequences of DNA, these assumptions greatly simplify problems in alignment. In this case

$$L(y_i) = \prod_{j=1}^{n_i} P(y_{ij})$$

and the only parameters entering the likelihood are the frequencies of observing the different sequence elements. But if we consider the joint likelihood

$$L(y_1, y_2) = L(y_1|y_2) \prod_{j=1}^{n_2} P(y_{2j})$$

we still must consider the conditional distribution of y_1 given y_2. Given our previous assumption that the elements of the 2 sequences arise in an iid fashion, it seems natural to assume that if f_1 and f_2 are the 2 functions that map the sequences to each other for a given alignment then (i.e. the 2 functions that stick the minimal number of gaps in the sequences so that we can then line them up),

$$L(y_1|y_2) = \prod_{j=1}^{m} P\Big(f_1(y_{1j})|f_2(y_{2j})\Big)$$

(where m is a parameter that depends on the number of gaps used in the alignment).
Note that this assumes that mutations at adjacent positions occur independently of
each other, and this is also potentially violated. These assumptions allow us to express the joint log-likelihood of our sequences as

$$\log L(y_1, y_2) = \sum_{j=1}^{m} \log\Big(P\Big(f_1(y_{1j})|f_2(y_{2j})\Big) P\Big(f_2(y_{2j})\Big)\Big).$$

If we tried to use the resulting likelihood, we would find that we need to estimate
$P(y_j = k)$ for all possible values of k, in addition to f_1 and f_2. Clearly these parameters are not identified with the given data since many combinations of these lead to
the same value of the likelihood. From a more practical perspective, we don't even
have any information on functional similarity of biosequences that we are using to estimate parameters. A pragmatic approach to these problems is to determine estimates
of

$$s(j, k) = \log\Big(\frac{P(y_{1i} = j, y_{2i} = k)}{P(y_{1i} = j)P(y_{2i} = k)}\Big)$$

for all possible j and k using a data set with information about protein functionality
as it relates to sequence similarity, then use these estimates to find f_1 and f_2 for
a given pair of sequences. This is the usual approach that is adopted in searching
databases for homologous sequences.

Collections of $s(j, k)$ for amino acids are referred to as *substitution matrices*. A number of these have been derived using expert knowledge based on actual protein sequences with known functionality. There are 2 basic varieties in common use: PAM
and BLOSUM. The PAM substitution matrices are based on an evolutionary model
whose parameters are estimated using closely related proteins. One then extrapo-
lates the observed amount of evolutionary divergence to find substitution matrices
for more distantly related proteins. In contrast, BLOSUM matrices determine how
much residues will differ as we require a certain amount of conservation between the
sequences. Each variety has a number of substitution matrices indexed by a value that
indicates the extent of divergence one expects between the 2 sequences. For closely
related sequences, one should use a PAM matrix with a low value, such as PAM30,
or a BLOSUM matrix with a high value, such as BLOSUM80. For the PAM matrices
the number indicates how much evolutionary time to allow between the sequences
(although the time unit does not admit a direct interpretation). For BLOSUM matrices the number indicates what percentage of residues is common across sequences.
Note that neither of these types of substitution matrices indicate what gap penalty
(i.e. the cost of aligning a residue in one sequence to a space in the other) should
be used. Recent empirical evidence for gap penalties when using the BLOSUM50
or PAM200 substitution matrices suggests a value of about 5 works well (Reese and

Table 7.1 *The BLOSUM50 substitution matrix.*

	A	R	N	D	C	Q	E	G	H	I	L	K	M	F	P	S	T	W	Y	V
A	5	-2	-1	-2	-1	-1	-1	0	-2	-1	-2	-1	-1	-3	-1	1	0	-3	-2	0
R	-2	7	-1	-2	-4	1	0	-3	0	-4	-3	3	-2	-3	-3	-1	-1	-3	-1	-3
N	-1	-1	7	2	-2	0	0	0	1	-3	-4	0	-2	-4	-2	1	0	-4	-2	-3
D	-2	-2	2	8	-4	0	2	-1	-1	-4	-4	-1	-4	-5	-1	0	-1	-5	-3	-4
C	-1	-4	-2	-4	13	-3	-3	-3	-3	-2	-2	-3	-2	-2	-4	-1	-1	-5	-3	-1
Q	-1	1	0	0	-3	7	2	-2	1	-3	-2	2	0	-4	-1	0	-1	-1	-1	-3
E	-1	0	0	2	-3	2	6	-3	0	-4	-3	1	-2	-3	-1	-1	-1	-3	-2	-3
G	0	-3	0	-1	-3	-2	-3	8	-2	-4	-4	-2	-3	-4	-2	0	-2	-3	-3	-4
H	-2	0	1	-1	-3	1	0	-2	10	-4	-3	0	-1	-1	-2	-1	-2	-3	2	-4
I	-1	-4	-3	-4	-2	-3	-4	-4	-4	5	2	-3	2	0	-3	-3	-1	-3	-1	4
L	-2	-3	-4	-4	-2	-2	-3	-4	-3	2	5	-3	3	1	-4	-3	-1	-2	-1	1
K	-1	3	0	-1	-3	2	1	-2	0	-3	-3	6	-2	-4	-1	0	-1	-3	-2	-3
M	-1	-2	-2	-4	-2	0	-2	-3	-1	2	3	-2	7	0	-3	-2	-1	-1	0	1
F	-3	-3	-4	-5	-2	-4	-3	-4	-1	0	1	-4	0	8	-4	-3	-2	1	4	-1
P	-1	-3	-2	-1	-4	-1	-1	-2	-2	-3	-4	-1	-3	-4	10	-1	-1	-4	-3	-3
S	1	-1	1	0	-1	0	-1	0	-1	-3	-3	0	-2	-3	-1	5	2	-4	-2	-2
T	0	-1	0	-1	-1	-1	-1	-2	-2	-1	-1	-1	-1	-2	-1	2	5	-3	-2	0
W	-3	-3	-4	-5	-5	-1	-3	-3	-3	-3	-2	-3	-1	1	-4	-4	-3	15	2	-3
Y	-2	-1	-2	-3	-3	-1	-2	-3	2	-1	-1	-2	0	4	-3	-2	-2	2	8	-1
V	0	-3	-3	-4	-1	-3	-3	-4	-4	4	1	-3	1	-1	-3	-2	0	-3	-1	5

Pearson, 2002). For short sequences, one should use a lower gap penalty since one will only find short alignments and these tend to give lower alignment scores. BLO-SUM substitution matrices are thought to outperform the PAM matrices for distantly related sequences. The BLOSUM50 substitution matrix is shown in Table 7.1.

7.4 Dynamic programming

Dynamic programming is a general technique that can be used to solve many sequential optimization problems, even with complicated constraints on the solutions. The technique is fundamental in several areas of probability and statistics such as stochastic control, optimal stopping and sequential testing. While many problems can be addressed with the technique, it has some fundamental limitations. To fix ideas, it is best to discuss the technique with an example in mind.

Suppose we want to traverse a rectangular lattice from the upper left corner to the lower right corner in a highest scoring fashion. We require that we can only move to an adjacent location on each move and each move has a certain score associated with it. Furthermore, we constrain the solution so that we only move down, right or diagonally down and to the right. We could find the highest scoring path by writing down every possible path, computing the score for each path and then selecting the highest scoring path. While this technique would find the optimal path, it could take a very long time if the lattice is large since there are many possible paths.

The principle of dynamic programming can be used to find the optimal path much more quickly. The principle of dynamic programming is as follows: if a path is an optimal path, then whatever the best initial first move is in this path, the rest of the path is optimal with respect to this first move. In other words, we don't know what the optimal path is, but whatever the optimal path is, if a location, x, is visited by that path, the rest of the optimal path is the best path of those paths that start at location x. The proof of the principle is by contradiction: suppose the principle is not true, then an optimal path has the property that it optimizes the total score of traversing the lattice, yet such a path is suboptimal if we start at the second location of the optimal path for traversing the entire lattice.

Returning to our example, this principle means that we don't have to check every path. Instead we go to each location in the lattice and ask: if this location is visited by the optimal path, what was the highest scoring way to get here and what is the score? We ask this question of every location in the lattice, starting at the upper left corner and working to the right and down. If we store all this information (i.e. the optimal way to get to each location and the score for doing so), then when we have asked the question of the location in the bottom right corner, we will have the score of the highest scoring path, and we will know what is the optimal move to get to the bottom right corner from the previous location in the optimal path. We can then go to this previous location in the optimal path and find out what was the optimal way to get there (since we stored all that information). We then continue the process of tracing back the previous optimal location until we reach the upper left corner. In the end, we have the highest scoring path and its score. If g is the gap penalty and F is a matrix with element $F(i,j)$ that holds the optimal score if the optimal path goes through location i, j then we recursively compute

$$F(i,j) = \max(F(i-i,j) + g, F(i,j-1) + g, F(i-1,j-1) + s(y_{1j}, y_{2j})),$$

and store which of $F(i-i,j) + g$, $F(i,j-1) + g$ or $s(y_{1j}, y_{2j})$ is the largest for all i and j.

7.5 Using dynamic programming to find the alignment

If we are given 2 sequences and the penalties for mismatches and gaps, we can use dynamic programming to find the optimal alignment. To this end, suppose sequence 1 is of length n_1 and sequence 2 is of length n_2. Consider the $n_1 + 1$ by $n_2 + 1$ lattice with each sequence written along one of the margins. An alignment can be represented as a path that only goes down and to the right from the upper left corner to the lower right corner. If the path moves to the right in the ith row then the ith residue in sequence 1 is aligned to a gap, if the path goes down in the ith column then the ith residue in sequence 2 is aligned to a gap. If the path goes diagonally, then the residues in the rows that are at the endpoint of the path are aligned to each other.

Every allowable alignment can be represented as one of the paths we have described above, hence since we are given the scores for mismatches we can determine the score for each path. Notice that the score for each path is just the sum of the scores

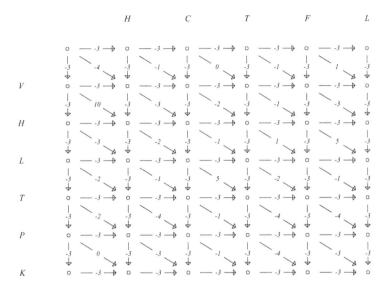

Figure 7.2 *The collection of possible choices for finding an alignment and the corresponding scores for the choices.*

from all of the alignments. Hence every connection between 2 adjacent locations has a score associated with it. This is exactly the sort of problem we introduced when we discussed dynamic programming, and so we can use dynamic programming to find the optimal alignment. This method is referred to as the Needleman Wunsch algorithm. The score for aligning 2 sequences given the mismatch scores and the gap penalties is referred to as the *alignment score*.

Consider aligning the following 2, unrealistically short, sequences: VHLTPK and HCTFL. One starts by writing each sequence along the margin of what will become the dynamic programming table, as in Figure 7.2. We will suppose that sequence 1 is put on the left margin and sequence 2 is put along the top of the table. Next one considers the array with dimension given by $n_1 + 1$ and $n_2 + 1$. One then fills in the scores associated with using each of the line segments. For example, horizontal and vertical moves require insertion of a gap, so the score associated with those links are all given by the gap penalty (here -3). One gets the score associated with the arrow that ends at position i, j for $i = 1, \ldots, n_1 + 1$ and $j = 1, \ldots, n_2 + 1$ by determining the score for matching the amino acids the $(i - 1)^{\text{th}}$ residue in sequence 1 to the residue in the $(j - 1)^{\text{th}}$ position in sequence 2. The result is shown in Figure 7.2. One then finds the optimal path by going to each dot and determining which of the at most 3 possible ways of getting there gives the best score. One then stores this in-

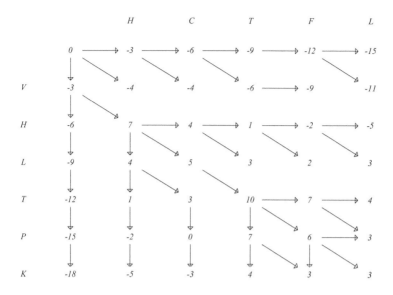

Figure 7.3 *The solution to the alignment problem using dynamic programming.*

formation for each location. Since there is only one path that goes through the upper and left margins (i.e long stretches of repeated gaps) these are easy to obtain and are represented by arrows pointing from which direction one should have moved to arrive at that location. One also stores the scores associated with visiting that location. The results are shown in Figure 7.3. Consider $F(1, 2)$: if location $(1, 2)$ is visited by the optimal path, then the optimal path must go from $(1, 1)$ to $(1, 2)$ and this would involve inserting a gap leading to a score given by the gap penalty. So we put the gap penalty in the location $(1, 2)$ and an arrow from $(1, 1)$ to $(1, 2)$. Now, if $(1, 3)$ is visited by the optimal path, then the path must go through $(1, 2)$, which would give a score given by the sum of the score to get to $(1, 2)$ plus the gap penalty, here -3-3=-6. This is then stored at that location along with the arrow giving that direction. The locations where $i > 1$ and $j > 1$ are more interesting since there are 3 possible paths to these internal sites. For example, if the optimal path visits $(2, 2)$ then there are 3 possible previous locations: $(1, 1)$, $(1, 2)$ and $(2, 1)$. The scores associated with visiting each of those sites is already stored in our table: 0, -3 and -3 respectively. Since the score for aligning a V with an H is -4, the total score of going from $(1, 1)$ to $(2, 2)$ is just 0-4=-4 and this is larger than either of the other 2 possible scores (both -6). We then proceed to fill in the table, ultimately getting the result displayed in Figure 7.3. Note that there is a tie at position $(6, 6)$. One then finds the optimal path by tracing back along the sequence of arrows starting in the lower right hand corner, here giving the optimal alignment of

```
VHLTPK
-HCTFL.
```

The size of the lattice largely determines how quickly the method will work. Recall the point of dynamic programming is that it is faster than enumerating all paths because we just query each location. Typically the number of paths grows exponentially with the average sequence length, whereas the number of locations only grows like the product of the 2 sequence lengths. But for long sequences, a computation depending on the product of their lengths can even become too computationally intensive, hence dynamic programming can become infeasible in some situations. On the other hand, it is very easy to program and is guaranteed to find the global optimum (or all of the global optima if there is more than one).

Returning to the IL7 example, we can use dynamic programming to find the optimal alignment. If we use the BLOSUM50 substitution matrix with a gap penalty of -4 we can find the optimal alignment. If we display the mouse protein on the top and use the dash symbol (i.e. −) to represent an insertion, we obtain the following alignment:

```
        10          20          30          40          50

MFHVSFRYIF GIPPLILVLL PVTSSECHIK DKEGKAYESV LMISIDE-LD

MFHVSFRYIF GLPPLILVLL PVASSDCDIE GKDGKQYESV LMVSIDQLLD

        60          70          80          90         100

KM--TGTDSN CPNNEPNFFR KHVCDDTKEA AFLNRAARKL KQFLKMNIS-

SMKEI--GSN CLNNEFNFFK RHICDANKEG MFLFRAARKL RQFLKM-NST

       110         120         130         140         150

EEFNVHLLTV SQGTQ-TLVNC T--S----- ---K-E--E- ---K--NVKE

GDFDLHLLKV SEG-TTILLNC TGQVKGRKP AALGEAQPTK SLEENKSLKE

       160         170         180

QK-KNDACFL KRLLREIKTCW NKILKGS-- I

QKKLNDLCFL KRLLQEIKTCW NKILMGTKE H.
```

Note the large insertion in the mouse version that we witnessed in the dot plot is also present in this alignment (from position 121 to 137). Also note the lack of a unique solution here. We could move the insertion in the mouse sequence at position 48 to position 49 and align the L in position 49 to the L in position 48 in the human sequence and obtain the same score. While this frequently occurs it doesn't alter the overall view of the level of similarity between 2 sequences.

7.5.1 Some variations

There are a number of simple refinements that can be made to increase the realism of the alignment method discussed above. For example, opening a gap may be the primary obstacle preventing a mutation from occurring, but once a gap has been opened it may be easy to extend it since having a gap there doesn't affect biological function. Hence we may want to allow different penalties for gaps depending on whether the gap is opened or extended. This is easy to accommodate within the context of dynamic programming because when we find the highest score for reaching each location we just need to consider if there was a gap leading to each of the previous locations. In contrast, if we had a penalty that depended on if there was a gap 2 residues ago, we would encounter much more bookkeeping and the dynamic programming approach becomes more computationally intensive. We can also have position dependent scores, for example, we may want to say to sequences are more similar if they have more residues in common in some region of the molecule. For example, a ligand binding site may be known for some protein, hence we only consider proteins to be matches if they have a lot of sequence in common in this region.

7.6 Global versus local alignments

Thus far we have been finding alignments over the entire sequence, yet typically biologists think alignment over short distances is what really matters for functional conservation in many instances. The alignments we have sought are known as *global alignments*, in contrast to aligning short segments of 2 given sequences, which is known as *local alignment*. Local alignment is more relevant because many proteins have functions that are mediated by their ability to bind to another molecule (the ligand of the protein), hence function will be preserved if this shorter segment is retained through evolution even if there is substantial divergence in other regions of the proteins. Since proteins are folded in their natural state, these preserved regions needn't actually be continuous segments of the protein. In fact, many researchers working on lymphocyte recognition of antigen (i.e. foreign molecules) explicitly account for these discontinuities in binding domains (so called "non-linear" epitopes, where an *epitope* is the ligand of a lymphocyte).

We can easily modify our global alignment algorithm to become a local alignment algorithm by introducing stops in our dynamic programming algorithm. That is, when we compute the maximal scores at each location we add a fourth option: have no alignment. This means instead of just comparing the 3 possible scores, we add another option in which the total score is simply zero. If $F(i, j)$ represents the value in the array of scores, g is the gap penalty, r_{ij} is the jth element of the ith sequence, then we now have

$$F(i, j) = \max(0, F(i - i, j) + g, F(i, j - 1) + g, F(i - 1, j - 1) + s(y_{1j}, y_{2j})),$$

where s is the score for 2 elements from the sequences. With this modification, one then proceeds to fill in the matrix $F(i, j)$ as before. To find the optimal local alignment using this matrix of scores, one first determines the maximal value in the dynamic programming table. This is where the optimal local alignment *ends*. One then traces back from this location to the first location which is greater than or equal to zero in the matrix F. This point is the start of the optimal local alignment. This algorithm is known as the *Smith Waterman algorithm*. For our previous example, we find that the best local alignment is

```
HLT
HCT.
```

7.7 Exercises

1. Find the optimal alignment of the 2 protein sequences ALTN and AFSDG using the BLOSOM50 score matrix and a gap penalty of 5.

2. If one compares 2 completely unrelated sequences of the same length, n, what proportion of the squares in a dot plot would be 1 if the sequences have elements from an alphabet of length a? Do you think it is easier to detect homology between proteins or DNA segments (of the same length)?

3. If one compares 2 completely unrelated nucleotide sequences of the same length, n where n is large, what is the average length of a diagonal line found in a dot plot?

4. Make a dot plot to compare the human hemoglobin subunit alpha, with sequence

```
         10         20         30         40         50
MVLSPADKTN VKAAWGKVGA HAGEYGAEAL ERMFLSFPTT KTYFPHFDLS
         60         70         80         90        100
HGSAQVKGHG KKVADALTNA VAHVDDMPNA LSALSDLHAH KLRVDPVNFK
        100        120        130        140
LLSHCLLVTL AAHLPAEFTP AVHASLDKFL ASVSTVLTSK YR
```

to the mouse version of the same protein, with the following sequence.

```
         10         20         30         40         50
MVLSGEDKSN IKAAWGKIGG HGAEYGAEAL ERMFASFPTT KTYFPHFDVS
         60         70         80         90        100
HGSAQVKGHG KKVADALASA AGHLDDLPGA LSALSDLHAH KLRVDPVNFK
        100        120        130        140
LLSHCLLVTL ASHHPADFTP AVHASLDKFL ASVSTVLTSK YR.    Hint: Use
```

the SWIS-PROT database to obtain these sequences.

5. Use the Needleman Wunsch algorithm to find the best alignment of the human and mouse version of the hemoglobin subunit alpha with sequences as given in the previous exercise (use the BLOSUM50 substitution matrix).

6. Use the Smith Waterman algorithm to find the best alignment of the human and mouse version of the hemoglobin subunit alpha with sequences as given in the previous exercise (use the BLOSUM50 substitution matrix).

7. Assuming that all amino acids occur with equal frequency, show how to simulate the evolution of a sequence of amino acids subject to only mutations using a given substitution matrix.

CHAPTER 8

Significance of Alignments and Alignment in Practice

8.1 Statistical significance of sequence similarity

The basis for sequence alignment is that there is a common ancestor for the sequences under investigation. We derive each of the 2 sequences under comparison by a sequence of mutations, insertions and deletions (insertions and deletions are referred to as *indels*). The process of obtaining a sequence from another by these basic operations works forward and backward in time, i.e. a deletion forward in time is an insertion if one reverses time. Hence, if y_0 is the common progenitor of the 2 sequences, y_1 and y_2, our evolutionary model postulates that $y_1 = M_s(M_{s-1}(\cdots(y_0)\cdots))$ and $y_2 = M_t(M_{t-1}(\cdots(y_0)\cdots))$, where M_s represents a mutation or indel. But since mutations and indels are time reversible, we can write

$$y_1 = f_J(f_{J-1}(\cdots(y_2)\cdots)),$$

for some sequence of mutations and indels, f_j. Assuming all f_j have the same impact on the function of the protein (which is not realistic but makes the statement of the problem most simple), the hypothesis that the 2 sequences are not functionally related is $H_0 : J \geq J_0$, where J_0 is large enough so that the protein has lost its function. Unfortunately, even with the simplifying assumption regarding equal impact on functionality for all f_j, little is known about J_0 and this approach is not employed in practice.

The practical approach to statistical significance adopted is to use data on proteins with known evolutionary similarity and use these to determine the likelihood of a given mutation. As discussed in Section 7.3, we can use expert knowledge to obtain estimates of how likely given mutations will impact function, and then use dynamic programming to find the most likely alignment and an alignment score. In fact, the alignment score is the log of the likelihood ratio for testing if the 2 sequences are independent against the alternative that they are related and the probabilities for mutations are as specified in the substitution matrix (and gap penalty) being used. The natural question then is if this alignment score is extreme enough to conclude the alignment is not just chance agreement. Unlike the usual applications of the theory of maximized likelihood ratio statistics, the number of parameters increases with the number of independent observations on the process (i.e. the length of the 2 sequences) since the alignment itself is a parameter that is being estimated from the data. This makes the usual theory of such tests not relevant, and indeed the distribution of the alignment score under the null is not like any χ^2 random variable.

Note that the optimal alignment is the highest of many possible scores, that is, if there are N distinct paths through the lattice introduced for finding the optimal alignment, and if s_i is the score associated with the ith path, then our alignment score is $S = \max_i s_i$. Since many of the paths necessarily intersect, the scores themselves, s_i, can not be independent, although this dependence is weak if the length of the sequences is long. In addition, we can model these scores as drawn from the common distribution defined by the sets of sums of gap penalties and alignment scores. Hence to determine the distribution of the alignment score under the null hypothesis, we need to consider the distribution of the maximum of a set of iid random variables.

8.2 Distributions of maxima of sets of iid random variables

In a paper of Gnedenko (1943) it was shown that the asymptotic distribution of the maximum of a set of iid variables has one of only 3 possible distributions. The asymptotic distribution of the maximum depends on the tail behavior of the cdf. We will consider only the 2 cases where the random variable is not bounded. In this case, if we let y_i for $i = 1, \ldots, n$ represent iid random variables then note that $\lim_{n \to \infty} \max_{1 \leq i \leq n} y_i = \infty$ since as more observations are collected there is always positive probability of observing a larger value of the random variable. For this reason, to get a large sample approximation we will attempt to find sequences $\{a_n\}$ and $\{b_n\}$ so that

$$\lim_{n \to \infty} \frac{\max_{1 \leq i \leq n} y_i - a_n}{b_n}$$

behaves like a random variable. This is directly analogous to considerations regarding the asymptotic distribution of the sample mean. To see this, suppose $y_i > 0$ for $i = 1, \ldots, n$ are iid random variables (with finite mean μ and variance σ^2). Then

$$\lim_{n \to \infty} \sum_{i=1}^{n} y_i = \infty$$

since

$$\lim_{n \to \infty} \frac{1}{n} \sum_{i=1}^{n} y = \mu$$

by the law of large numbers and the fact that the y_i are positive. In fact, the law of large number implies that

$$\lim_{n \to \infty} \left(\frac{1}{n} \sum_{i=1}^{n} y_i - \mu \right) / \sigma = 0.$$

One way to think about the central limit theorem is that we are trying to find some sequence $\{c_n\}$ so that

$$\lim_{n \to \infty} c_n \left(\frac{1}{n} \sum_{i=1}^{n} y_i - \mu \right) / \sigma$$

behaves like a random variable. If we can find such a sequence then we can make approximate statements about how close the sample mean is to the population mean μ, and this has many applications in statistical inference. The answer provided by the central limit theorem is that we should choose $c_n = \sqrt{n}$, and then

$$c_n \left(\frac{1}{n} \sum_{i=1}^{n} y_i - \mu \right) / \sigma$$

will behave like a $N(0, 1)$ random variable. This is the justification for the usual large sample approximation for the sample mean that states that $\bar{y} \sim N(\mu, \sigma^2/n)$. If we choose $c_n > \sqrt{n}$, say $c_n = n$, then $c_n(\frac{1}{n}\sum_{i=1}^{n} y_i - \mu)/\sigma$ will diverge, whereas if we choose $c_n < \sqrt{n}$, say $c_n = n^{\frac{1}{4}}$, then $c_n(\frac{1}{n}\sum_{i=1}^{n} y_i - \mu)/\sigma$ will go to 0 (these follow from the fact that selecting $c_n = \sqrt{n}$ results in the ratio acting like a random variable).

We can investigate this issue through some simulations. We can select a sample size, simulate a data set with this sample size, then find the sample maximum. If we repeat this exercise many times, say 1000, then we will have 1000 maxima. If we then find the mean (or median, here it doesn't matter in terms of the shape of the relationship depicted) of these numbers, we will have a Monte Carlo estimate of the theoretical mean of the sample maximum as it depends on the sample size. Figure 8.1 displays the result using the t-distribution with 1 degree of freedom and the standard normal distribution. For the t-distribution, we plot the maximum as it depends on the sample size, whereas for the normal distribution, we plot the maximum as it depends on the logarithm of the sample size. Note that for the t-distribution, the maximum increases like the sample size, but for the normal distribution, the maximum increases like the logarithm of the sample size. For densities with unbounded domains (i.e. random variables that can take an arbitrarily large values, like the t-distribution and the normal distribution), these are the only possible ways that the maximum can depend on the sample size.

To find the 2 sequences of $\{a_n\}$ and $\{b_n\}$ we will consider approximating the cumulative distribution function (cdf) of the maximum. Note that the cdf of the maximum of set of iid variables with cdf $F(x)$ is $F(x)^n$ since

$$P(\max_{1 \le i \le n} y_i \le x) = P(y_i \le x \text{ for all } i)$$

but since the y_i are independent and have the same cdf

$$\begin{aligned} P(\max_{1 \le i \le n} y_i \le x) &= \prod_{i=1}^{n} F(x) \\ &= F(x)^n \end{aligned}$$

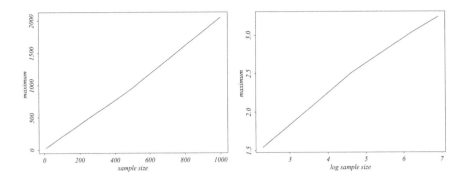

Figure 8.1 *Comparison of the relationship between the sample maximum and the sample size (or log sample size) for the t-distribution with 1 degree of freedom (left panel) and for the standard normal distribution (right panel). Here, the mean of 1,000 samples is shown as it depends on 4 different sample sizes: 10, 100, 500 and 1000.*

Let

$$Z = \max_{1 \leq i \leq n} y_i.$$

The first case we will consider is where F, the cdf of the iid observations, is such that $1 - F(y) \sim ay^{-c}$ for some a, c (both positive). In this case we can set $a_n = 0$. Then for any sequence b_n

$$\begin{aligned}
P(Z/b_n \leq z) &= \left[F(b_n z)\right]^n \\
&= \left[1 - \left(1 - F(b_n z)\right)\right]^n \\
&\sim \left[1 - a(b_n z)^{-c}\right]^n.
\end{aligned}$$

So letting $b_n = (an)^{1/c}$, we find

$$P(Z/b_n \leq z) \sim e^{-z^{-c}}.$$

If $1 - F$ goes to zero exponentially, i.e. $1 - F(y) \sim e^{-y}$ then we obtain a different distribution for the maximum. In this case

$$\begin{aligned}
P\left(\frac{Z - a_n}{b_n} \leq z\right) &= \left[F(a_n + b_n z)\right]^n \\
&= \left[1 - \left(1 - F(a_n + b_n z)\right)\right]^n \\
&= \left[1 - \exp\left\{\log\left(1 - F(a_n + b_n z)\right)\right\}\right]^n.
\end{aligned}$$

In this case we expect a_n to be near $F^{-1}(1)$, and we will set $a_n = F^{-1}(1 - 1/n)$.

Let
$$g(z) = \log(1 - F(a_n + b_n z)).$$
If we then find a first order Taylor expansion for g near $z = 0$ (using $a_n = F^{-1}(1 - 1/n)$) we find
$$g(z) \approx \log(n^{-1}) - nb_n f(a_n)z,$$
where f is the density of the random variables, so that
$$P\left(\frac{Z - a_n}{b_n} \le z\right) \approx \left[1 - \frac{1}{n}\exp\left\{-nb_n f(a_n)z\right\}\right]^n$$
and so if we set $b_n = 1/(nf(a_n))$ we find that
$$P\left(\frac{Z - a_n}{b_n} \le z\right) \sim \left[1 - \frac{1}{n}\exp\{-z\}\right]^n.$$
But then taking the limit in n we find
$$P\left(\frac{Z - a_n}{b_n} \le z\right) \sim \exp\{-e^{-z}\}. \tag{8.1}$$

This distribution is known as the *extreme value* distribution. Note that the above argument implies $EZ \sim a_n$, but $a_n \sim \log n$, so $EZ \sim \log n$, which agrees with our simulation result in Figure 8.1. Note also that since $1 - F(y) \sim e^{-y}$, $f(y) \sim e^{-y}$ so that $f(a_n) \sim n^{-1}$. This implies that if we set $b_n = 1/(nf(a_n))$ then $a_n/b_n \sim c + \log n$ for some constant c. If we then combine this with equation 8.1, we find that
$$P(Z \le z) \sim \exp\left\{-\kappa n e^{-\lambda z}\right\} \tag{8.2}$$
for 2 parameters κ and λ.

8.2.1 Application to sequence alignment

The usual approach to determining the statistical significance of an alignment supposes the tail area of the distribution of alignment scores decays at an exponential rate, hence the extreme value distribution is used to determine tail areas. This is appropriate since the scores are sums and provided the correlation within a sequence isn't too strong a central limit theorem implies the convergence of these sums to a normal distribution. When we are finding the local alignment, if we let n and m represent the lengths of the 2 sequences being aligned, the size of the largest local alignment will depend on the product of the length of the sequences nm (since there are nm possible starting locations for an optimal alignment), hence the sequence alignment score, S, is such that $ES \approx \log nm + \text{constant}$ or $ES \approx \log \kappa nm$ for some κ. For this reason, the extreme value distribution is parameterized as $\exp\{-\kappa nm e^{-\lambda S}\}$ for 2 parameters κ and λ (here S is the alignment score) as in equation 8.2. We could compute the p-value associated with an alignment score of $s_{observed}$ by the expression
$$P(S \ge s_{observed}) \approx 1 - \exp\{-\kappa nm e^{-\lambda s_{observed}}\}$$

if we knew κ and λ. While one can obtain expressions for κ and λ in terms of the substitution matrix used when there are no gaps (see Karlin and Altschul, 1990), no such results are available for gapped alignments. However, a number of researchers have estimated these parameters using a variety of proteins that are unrelated (see Altschul and Gish, 1996; Reese and Pearson, 2002). This approach provides reasonable approximations to the values of these parameters under the null and the resulting p-values are referred to as e-values.

8.3 Rapid methods of sequence alignment

Often a researcher has a new protein or DNA sequence and wants to find the sequences in a large database that are similar to the new sequence. When the sequences are 1000's of base pairs long, dynamic programming will take a small but noticeable amount of time to find the optimal alignment. Thus if we are searching through a database with millions of sequences, dynamic programming can become impractical. For this reason, a number of generally suboptimal but rapid methods for sequence alignment are in use.

Recall from the dynamic programming formulation that if 2 sequences have extensive agreement, the optimal path in the dynamic programming table will be largely diagonal. Hence a quick way to gauge the local similarity of 2 sequences is to count the length of the longest diagonal in the dot matrix form. While it would seem summing the diagonals of a matrix would take on order nm (i.e. there are Cnm calculations necessary to perform the task for some constant C), there are clever algorithms for computing these sums more rapidly. This is the basic trick to the 2 most popular, *ad hoc*, rapid methods for sequence alignment, BLAST and FASTA.

The simple way to find the diagonal sums in the dot plot would involve examining every position in the dot plot, and this requires visiting every position and determining if there is a dot or not. Since the number of distinct diagonal sums is the sum of the lengths of the 2 sequences minus 1 and the majority of the positions are not dots, a better way would involve just counting these diagonal sums and all the locations that have dots in them. To illustrate the method, suppose we want to align the 2 sequences

ATCAATTA

and

AATGCAATG.

To find the alignment, one first performs an operation on one of the 2 sequences to be aligned that is called *hashing* the k-tuples, and the results of this operation are stored in a *hash table*. A k-tuple is a subsequence of contiguous elements of a given sequence of length k. The process of hashing is the process of finding all k-tuples of some sequence and noting the start locations of these subsequences. This is of interest in itself as we can easily find repeated subsequences within the original sequence. For example, we can make a table that lists all possible 2-tuples and gives their start locations, which for the second sequence is given in Table 8.1.

Table 8.1 *A hash table for the sequence* AATGCAATG.

2-tuple	start positions
AA	1, 6
AT	2, 7
CA	5
GC	4
TG	3, 8

We can then use this table to align the 2 sequences as follows. The basic idea is that we will construct all the diagonal sums at the same time that have at least a 2-tuple in them. We will then obtain all locations that have a 2-tuple in common across both sequences. If we then repeat this process for 3-tuples, 4-tuples and so on, we will eventually discover that the largest common subsequence in the 2 sequences is the 4-tuple CAAT. While it would seem to be time consuming to consider all k-tuples, the computational time for each is greatly reduced compared to the naive method of summing the diagonals. This reduction is by a factor of $(\sum_i p_i^2)^k$, where p_i is the probability of observing the occurrence of the ith possible sequence element (see Wilbur and Lipman, 1983). For DNA, if we suppose all nucleotides are equally common and let $k = 6$, we find this factor is about $2.0e^{-4}$, while if we suppose all amino acids are equally common and let $k = 2$ we find that the factor is about $3.0e^{-3}$. Additionally, as we will see below, clever uses of this idea don't consider every possible k-tuple.

To make the discussion more general while illustrating how the algorithm works with this example, we will suppose that the sequence we have constructed the hash table for is of length m while the other sequence is of length n. Hence there are $n + m - 1$ distinct diagonals, where 2 of these diagonals are of length 1. Thus the total number of diagonals that contain at least a 2-tuple are $n + m - 3$ (or $n + m - 2k + 1$ for an arbitrary k-tuple where k is much smaller than n and m). Define the matrix representation, A (with element a_{ij}), of the dot plot to be a matrix of 0's and 1's where $a_{ij} = 1$ if the ith element of sequence 1 is the same as the jth element of sequence 2. Note that the indices of the matrix representation of the dot matrix have the property that if one considers the diagonals that run down and to the right, then the difference in the indices is the same for all elements belonging to the same diagonal. For example, the elements along some of the diagonals of the matrix A are $a_{n,1}$ (where $i - j = n - 1$), $(a_{n-1,1}, a_{n,2})$ (where $i - j = n - 2$), $(a_{1,m-1}, a_{2,m})$ (where $i - j = 2 - m$), and $a_{1,m}$ (where $i - j = 1 - m$). For this reason we will consider the diagonal sums S_d where $d = i - j$ for $k - m \leq d \leq n - k$, and we will build these sums one element at a time assuming initially that $S(d) = 0$ for all d. We begin by scanning each 2-tuple in the first sequence with reference to the hash table for the second. For example, the first 2-tuple in the first sequence is AT. Looking at the hash table we see that sequence 2 has 2 instances of AT, at locations 2 and 7. In this case $i = 1$ and for the other sequence $j = 2$ and $j = 7$ are matches, thus

Table 8.2 *The use of a hash table for finding 2-tuples common across 2 sequences.*

iteration (i)	position of matches (j)	$i - j$	altered diagonal sums
1	2,7	-1,-6	$S_{-1} = 1, S_{-6} = 1$
2	-	-	-
3	5	-2	$S_{-2} = 1$
4	1,6	3,-2	$S_{-2} = 2, S_3 = 1$
5	2,7	3,-2	$S_{-2} = 3, S_3 = 2$
6	-	-	-
7	-	-	-

we increment the sums S_{-1} and S_{-6} by 1. The next 2-tuple in the first sequence is TC, but the hash table indicates that there are no TC's in the second sequence, so no diagonal sums are incremented. For the third 2-tuple, $i = 3$, we find a match at $j = 5$ so we increment S_{-2} by one. After finishing the the rest of the iterations (over i), as summarized in Table 8.2, we find that $S_{-6} = 1$, $S_{-2} = 3$, $S_{-1} = 1$, and $S_3 = 2$ are the only non-zero diagonal sums when counting 2-tuples. This is easily confirmed by constructing a dot plot.

8.3.1 FASTA

FASTA is the latest version of a software package that is widely used for sequence alignment. It includes methods for aligning amino acid sequences and nucleotide sequences. The rapid heuristic search method is composed of several stages. At first the algorithm identifies long diagonal stretches in the dot plot. The length of these stretches is controlled by a parameter called ktup (which is short for k-tuple). As we expect based on the discussion of searching for k-tuples, small values of ktup allow for finding more stretches, whereas larger values require a greater length for the match (ktup is frequently set to 2 for protein searches and 4-6 for nucleotide searches). The algorithm then attempts to combine these local diagonal stretches by penalizing the combined regions depending on how many gaps are introduced. One then rescans these alignments using the scoring method of choice (i.e. a substitution matrix and a gap penalty) and removes ends that do not contribute to the alignment. Following this, there is another attempt to combine distinct subregions by introducing gaps and penalizing the scores of the combined alignments. Finally, a banded Smith Waterman algorithm is run (i.e. a band is set up around the best alignment found thus far and the Smith Waterman algorithm is run within that band) to find the optimal local alignment in the region identified using the heuristic search technique.

8.3.2 BLAST

Another popular heuristic algorithm for pairwise is known as BLAST (which stands for basic local alignment search tool). This algorithm proceeds by looking for short matches using the k-tuple approach described above (by default, k is set to 3 for protein searches and 11 for nucleic acids), known as seeds, and then tries to extend these seeds out in both directions. Scores are computed as the seeds are extended and those not passing a threshold are dropped from further consideration. There are constant refinements to this method and details of past implementations can be found in Altschul et al. (1990) and Altschul et al. (1997). The best way to obtain up to date information regarding details of the best performing version of these algorithms is to check on-line (by searching BLAST). Currently the National Center for Biotechnology Information (NCBI) allows on-line use of the state of the art implementation in conjunction with access to the most recently updated databases of sequence information (which are described in Section 8.4).

To use these search procedures, one needs to prepare a file that conforms to the FASTA format. This format has a first line that starts with a > symbol which is then followed by a unique number to identify the sequence (the *accession number* which is discussed in Section 8.4) and information about the sequence (these things are optional). The next line specifies the nucleotide or amino acid sequence in terms of the one letter codes. These should be plain text files. BLAST can be used to find an alignment of the mouse and human versions of the IL-7 protein we examined in section 7.5. To do so, one searches for BLAST on the web, then selects protein BLAST (BLASTp) from the BLAST homepage. There is then an option to just align 2 sequences (rather than to do database searching as is the usual application), hence if one selects this and enters the human and mouse sequences into the box provided in FASTA format, then clicks on the blast button, one obtains output that describes the alignment. This description includes some basic information on the alignment, such as the e-value, the proportion of residues that are identical after being aligned, the proportion of residues that give a positive score using the alignment (which is larger than the proportion that are identical since some non-identical residues have positive scores) and the proportion of gaps. The results also display the optimal alignment using the substitution matrix and gap penalty that are specified. In addition the gapped subsequence which is common along with positive scoring mismatches is displayed between the 2 aligned sequences. Here is the result from our example:

```
>lcl|32027 unnamed protein product

Length=154

Score = 209 bits (532), Expect = 2e-59,

    Method: Compositional matrix adjust.

Identities = 107/174 (61%), Positives = 128/174 (73%),

    Gaps = 21/174 (12%)

Query  MFHVSFRYIFGLPPLILVLLPVASSDCDIEGKDGKQYESVLMVSIDQ

       MFHVSFRYIFG+PPLILVLLPV SS+C I+ K+GK YESVLM+SID+

Sbjct  MFHVSFRYIFGIPPLILVLLPVTSSECHIKDKEGKAYESVLMISIDE
```

```
Query  LLDSMKEIGSNCLNNEFNFFKRHICDANKEGMFLFRAARKLRQFLKM

       L D M  SNC NNE NFF++H+CD KE FL RAARKL+QFLKM

Sbjct  L-DKMTGTDSNCPNNEPNFFRKHVCDDTKEAAFLNRAARKLKQFLKM

Query  NSTGDFDLHLLKVSEGTTILLNCTGQVKGRKPAALGEAQPTKSLEEN

       N + +F++HLL VS+GT  L+NCT +                    E

Sbjct  NISEEFNVHLLTVSQGTQTLVNCTSK------------------EE

Query  KSLKEQKKLNDLCFLKRLLQEIKTCWNKILMGT

       K++KEQKK ND CFLKRLL+EIKTCWNKIL G+

Sbjct  KNVKEQKK-NDACFLKRLLREIKTCWNKILKGS
```

While this alignment differs slightly from the optimal alignment found previously, both show a high level of homology and the large gap in the mouse sequence as compared to the human sequence. This used the default parameter settings, which at this time are the BLOSUM62 scoring matrix and a gap penalty that is 11 for opening a gap and 1 for extending an already existing gap. The program also reports an e-value, which here is $2.0e^{-59}$ indicating more agreement than one would expect by chance. Typically, closely related sequences have e-values less than $1.0e^{-50}$.

8.4 Internet resources for computational biology

There are many resources for computational molecular biology available on the internet (the book of Baxevanis and Ouellette, 2001, provides a detailed account of these resources). The first place to look is at the web site of the NCBI. This site supports BLAST searches (just click on links that say BLAST), interacts with many important databases and has access to many other interesting tools. The NCBI page is constantly changing, but it is a searchable site. Information on sequences is stored by accession numbers, which are identifiers assigned to each sequence in a database. If one knows the accession number, one can search for information on the sequence. For example, if you enter the number AF102503, then, if you search the EST records (or just search all databases) one obtains a search result (under EST in the resulting table if you search all databases). If you follow the link then you get a page full of information about that sequence, including the nucleotide sequence. You will see that someone from the University of Minnesota submitted this entry, it is an EST and some other information. If one goes to the bottom of the screen then there is a link to an article in *Virology* from which one could find out more abut this sequence. Other identification numbers are used for more complex objects than sequences. For example another commonly used identifier is the protein databank identifier, which is an identifier assigned by the protein databank (PDB), a database of 3-dimensional structures of nucleotide and amino acid sequences (obtained through X-ray crystallography or Nuclear Magnetic Resonance Imaging). Another example is the set of identifiers used in the Online Mendelian Inheritance in Man (OMIM), an NCBI linked index of most Mendelian genetic disorders with text descriptions and pictures.

In addition, one can access many databases from this site, a process coordinated by a search and retrieval tool called Entrez. For example, to obtain the DNA sequence for some gene or chromosome, one can examine the linked database called Genbank. Genbank is an annotated database of all known DNA sequences. This database is part of the International Nucleotide Sequence database Collaboration which entails comparing the 3 major DNA sequence databases, i.e. Genbank, the DNA Databank of Japan (DDBJ) and the European Molecular Biology Laboratory (EMBL), on a daily basis. For this reason, all of the major databases are nearly identical. Other interesting tools include programs that allow one to view 3-dimensional structures of proteins (i.e. the Molecular Modeling Database, MMDB) and the ability to find information on different types of genetic markers. For example, one can search the Marshfeild map (a commonly used genetic map) within the UniSTS database to determine on which chromosome the marker (i.e. D2S2924) that was linked to the quantitative trait developed in section 6.6 resides. Finally, there are many gene expression data sets (discussed more fully in the last 4 chapters) that one can access via the gene expression omnibus (GEO), and these are all annotated so that there is sufficient information to replicate the findings in the original publication (when searching for such a dataset using the accession number one can specify that the search should just be in the GEO). The sites of DDJB and EMBL also have many interesting links and programs.

There are also sites dedicated exclusively to protein data. For example, the UniProt database (with address www.uniprot.org) has access to many protein sequences, and these are highly curated, non-redundant and highly cross-linked to other databases. This database is the combination of 2 older databases, one of which is called Swiss-Prot and the other TrEMBLE (the second of these is not reviewed whereas the first was). As an example of using this database, if one searches the protein knowledgebase (UniprotKB) for information on the protein IL7, one finds many (over 180 currently) results where one is the human version of the protein (with accession number P13232). If one follows this link, then one can obtain the amino acid sequence. This was the source for the sequences used as an example in Section 7.2. This site changes and has useful features such as the ability to do BLAST searches. The previously mentioned Gene Ontology project also maintains a database that holds information regarding the ontology. All of these databases, and many others, are freely accessible for data submission and retrieval. Finally, there are also many databases that are designed for more specialized purposes that one would really only use in a specialized context. For example, there is a human mitochondrial protein database.

8.5 Exercises

1. Find the distribution of the maxima of a set of independent uniform on $(0,1)$ random variables (*Hint:* Consider $P\left(\frac{1-Z}{b_n} \leq x\right)$).
2. Simulate 100 draws from a standard normal distribution and find the maximum.

Repeat this exercise 1000 times so that you have 1000 samples from the distribution of the maximum of 100 draws from a standard normal distribution. Use maximum likelihood to estimate the parameters that appear in the general form of the extreme value distribution, and produce a graph that overlays the resulting density on a histogram of the 1000 maxima that were simulated.

3. Compare the amino acid sequence for IL7 for humans to that of mouse using a dot plot and locally align them using dynamic programming.

4. Construct the dot plot for the example where hashing was used to find 2-tuples. Show that the results are consistent with the dot plot.

5. Show the steps for using a hash table to find all 2-tuples that are common across across the 2 sequences ATTCTGA and ATTCGAAT.

6. Find the amino acid sequences for CD3 from a pig and a chicken (if there are multiple variants in the database, use the most recently annotate entries) and align the sequences using BLAST. Note the accession numbers of the entries that you align. Make a dot plot for these 2 sequences and compare alignment to the dot plot.

CHAPTER 9

Hidden Markov Models

9.1 Statistical inference for discrete parameter finite state space Markov chains

In Section 2.2.2 the idea of a Markov chain was introduced. Markov chains provide useful models for the analysis of sequence data since we can model the dependence that is found in biopolymers in a straightforward manner. To do so, we must consider the problem of parameter estimation for Markov chains. After addressing this problem, we will treat parameter estimation for hidden Markov models, which are widely used tools for the analysis of sequences of random variables that display dependence. Throughout this discussion we will suppose that the parameter that indexes the Markov chain is location on a biopolymer, hence we will refer to successive values as adjacent locations.

We can approach inference for Markov chains with the techniques of maximum likelihood or through Bayesian approaches. In either event, we must first consider how to compute the likelihood. Let θ denote a vector that holds the parameters involved in the model, i.e. the parameters in the initial distribution and the transition matrix. We first write the likelihood as the product of the conditional likelihoods, as we have seen before, using the Markov property

$$L(y_1, \ldots, y_n | \theta) = L(y_1 | \theta) \prod_{t=2}^{T} L(y_t | y_{t-1}, \theta).$$

Often the factor $L(x_1 | \theta)$ is just ignored since interest is usually on the parameters in the transition matrix. (Alternatively, we may specify a prior distribution for the beginning state.) The factors $L(y_t | y_{t-1}, \theta)$ will represent observations on a multinomial observation since y_t takes only finitely many values. The success probabilities for this multinomial observation will come from the row of P that corresponds to the value of y_{t-1}. As analogy with the binomial case, the MLEs for these multinomial probabilities will just be the sample proportions. Thus the MLE of the transition matrix is found by determining the sample proportions for each of the possible moves, i.e. the MLE of the i, j element of P is just the proportion of times that the chain moved from state i to state j. If there is prior information about the elements of P these can be incorporated through a prior that multiplies the likelihood in the usual fashion. Since we have an expression for the likelihood, the usual techniques of hypothesis testing and confidence interval construction are at our disposal (we just need to compute the observed information). Alternatively we can specify a prior for the elements of P, such as a set of Dirichlet priors (one for each row of P), and simulate values of these probabilities from their posterior.

9.2 Hidden Markov models

The fundamental idea behind a hidden Markov model is that there is a Markov process (i.e. a stochastic process with the Markov property) we cannot observe that determines the probability distribution for what we do observe. Thus a hidden Markov model is specified by the transition density of the Markov chain and the probability laws that govern what we observe given the state of the Markov chain. Given such a model, we want to estimate any parameters that occur in the model. We would also like to determine what is the most likely sequence for the hidden process. Finally we may want the probability distribution for the hidden states at every location. We now turn to each of these problems.

First we introduce some notation. Let y_t represent the observed value of the process at location t for $t = 1, \ldots, T$, θ_t the value of the hidden process at location t and let ϕ represent parameters necessary to determine the probability distributions for y_t given θ_t and θ_t given θ_{t-1}. In our applications, y_t will either be an amino acid or nucleotide and the hidden process will determine the probability distribution of observing different letters. Our model is then described by the sets of probability distributions $p(y_t|\theta_t, \phi)$ and $p(\theta_t|\theta_{t-1}, \phi)$ (in addition to an initial distribution $p(\theta_1|\phi)$). A crucial component of this model is that the y_t are independent given the set of θ_t and θ only depends directly on its neighbors θ_{t-1} and θ_{t+1}.

The various distributions in which we are interested are $p(\phi|y_1, \ldots, y_T)$ (for parameter estimation), $p(\theta_t|y_1, \ldots, y_T)$ for all t (for inference about the hidden process at a given location) and $p(\theta_1, \ldots, \theta_T|y_1, \ldots, y_T)$ (to determine the most likely path of the process). We will adopt a Bayesian perspective, so that we treat θ_t as a random variable. For the purposes of the next 2 sections, all probabilities are implicitly conditional on ϕ, i.e. we treat these as known. Later we will discuss estimation of ϕ. Much of the treatment here is based on Kitigawa (1987).

9.2.1 A simple binomial example

As a simple example to illustrate the ideas presented thus far, we will suppose that there is a hidden process θ_t that is in either of 2 states. We will denote these 2 states as 0 and 1 (for example, a DNA coding region and a noncoding region). Next we will assume that $y_t|\theta_t \sim \text{Bin}(n, f(\theta_t))$ where $f(0) = \phi_1$ and $f(1) = \phi_2$ for $\phi_1 < \phi_2$ (where the inequality is so that the model is identified). Thus, if we find that

$$p(y_t|\theta_t) = \binom{n}{y_t} \left[\phi_2^{y_t}(1 - \phi_2)^{n-y_t}\right]^{\theta_t} \left[\phi_1^{y_t}(1 - \phi_1)^{n-y_t}\right]^{1-\theta_t}.$$

We also specify that $p(\theta_t = 1 | \theta_{t-1} = 0) = \phi_3$ and $p(\theta_t = 1 | \theta_{t-1} = 1) = \phi_4$, so that we have the following transition matrix for the hidden Markov chain

$$\begin{pmatrix} 1 - \phi_3 & \phi_3 \\ 1 - \phi_4 & \phi_4 \end{pmatrix}.$$

This model has 4 unknown parameters $\phi = (\phi_1, \ldots, \phi_4)$ in addition to the sequence of hidden states θ_t. The usual goal is to conduct inference for θ_t for all t, but this is not possible without estimating ϕ too.

9.3 Estimation for hidden Markov models

9.3.1 The forward recursion

We first treat the problem of finding $p(\theta_t | y_1, \ldots, y_t)$ for all t. This will be an important component to finding the desired probability distributions, and in the state space models literature is known as *filtering*. Now, by applying Bayes theorem to θ_t and y_t, and then using the law of total probability we find

$$\begin{aligned} p(\theta_t = i | y_1, \ldots, y_t) &\propto p(y_t | \theta_t = i, y_1, \ldots, y_{t-1}) p(\theta_t = i | y_1, \ldots, y_{t-1}) \\ &\propto p(y_t | \theta_t = i) \sum_j p(\theta_t = i, \theta_{t-1} = j | y_1, \ldots, y_{t-1}), \end{aligned}$$

where $p(y_t | \theta_t = i, y_1, \ldots, y_{t-1}) = p(y_t | \theta_t = i)$ since y_t is independent of y_s for $s < t$ given θ_t. If we then use the definition of conditional probability in conjunction with the fact that θ_t is independent of $y_1, \ldots, y_{t-1}$ gievn θ_{t-1} we find

$$\begin{aligned} p(\theta_t = i | y_1, \ldots, y_t) &\propto p(y_t | \theta_t = i) \sum_j p(\theta_t = i | \theta_{t-1} = j, y_1, \ldots, y_{t-1}) \\ &\times p(\theta_{t-1} = j | y_1, \ldots, y_{t-1}) \\ &\propto p(y_t | \theta_t = i) \sum_j p(\theta_t = i | \theta_{t-1} = j) \\ &\times p(\theta_{t-1} = j | y_1, \ldots, y_{t-1}). \end{aligned}$$

So we have a recursion for the quantity $p(\theta_t = i | y_1, \ldots, y_t)$, hence we can update this quantity as we move across locations by using the 2 distributions that specify the model, $p(y_t | \theta_t)$ and $p(\theta_t | \theta_{t-1})$, and performing the necessary summation over j. To initiate the recursion, we typically specify a prior distribution for θ_1. For example we may suppose that θ_1 is equally likely to be in each possible state, or perhaps we have some idea about the frequency with which θ_1 would be in each possible state. In this case we have

$$p(\theta_1 = i | y_1) \propto p(y_1 | \theta_1 = i) p(\theta_1 = i).$$

By computing something to proportional to $p(\theta_1 = i | y_1)$ for all i, which we will denote q_i, we then find

$$p(\theta_1 = i | y_1) = \frac{q_i}{\sum_j q_j}$$

since $q_i = Cp(\theta_1 = i|y_1)$ for some constant C (common across all i). We can then use these probabilities to recursively find $p(\theta_t = i|y_1, \ldots, y_t)$ for all t. This recursion is often referred to as the *forward recursion* in the sequence analysis literature and *filtering* in other areas.

The forward recursion for the binomial example

We now return to the example first discussed in Section 9.2.1 to illustrate how one can use the forward recursion. First, suppose we have already computed $p(\theta_{t-1} = 0|y_1, \ldots, y_{t-1})$ and $p(\theta_{t-1} = 1|y_1, \ldots, y_{t-1})$, which we will denote $p_{0,t}$ and $p_{1,t}$ respectively, and stored these values (in fact only one of these needs to be stored since they sum to one). Then we have (ignoring the expression involving factorials since we are only finding something proportional to the posterior)

$$
\begin{aligned}
p(\theta_t = 0|y_1, \ldots, y_t) \quad &\propto \quad \phi_1^{y_t}(1 - \phi_1)^{n - y_t}\big[p(\theta_t = 0|\theta_{t-1} = 0)p_{0,t-1} \\
&+ \quad p(\theta_t = 0|\theta_{t-1} = 1)p_{1,t-1}\big] \\
&\propto \quad \phi_1^{y_t}(1 - \phi_1)^{n - y_t}\big[(1 - \phi_3)p_{0,t-1} \\
&+ \quad (1 - \phi_4)p_{1,t-1}\big].
\end{aligned} \tag{9.1}
$$

Similarly

$$
\begin{aligned}
p(\theta_t = 1|y_1, \ldots, y_t) \quad &\propto \quad \phi_2^{y_t}(1 - \phi_2)^{n - y_t}\big[p(\theta_t = 1|\theta_{t-1} = 0)p_{0,t-1} \\
&+ \quad p(\theta_t = 1|\theta_{t-1} = 1)p_{1,t-1}\big] \\
&\propto \quad \phi_2^{y_t}(1 - \phi_2)^{n - y_t}[\phi_3 p_{0,t-1} + \phi_4 p_{1,t-1}].
\end{aligned} \tag{9.2}
$$

If we denote the expressions on the right hand side of equations 9.1 and 9.2 as q_{0t} and q_{1t} respectively, then we obtain $p_{0t} = q_{0t}/(q_{0t} + q_{1t})$ and $p_{1t} = q_{1t}/(q_{0t} + q_{1t})$.

9.3.2 The backward recursion

Now we will suppose we have run through all of the data using our forward recursion, and we will use these results to obtain the posterior distribution of θ_t given all of the observations up to location y_T. To this end we can use the law of total probability along with the definition of conditional probability to find

$$
\begin{aligned}
p(\theta_t = i|y_1, \ldots, y_T) \quad &= \quad \sum_j p(\theta_t = i, \theta_{t+1} = j|y_1, \ldots, y_T) \\
&= \quad \sum_j p(\theta_t = i|\theta_{t+1} = j, y_1, \ldots, y_T) \\
&\quad \times \quad p(\theta_{t+1} = j|y_1, \ldots, y_T) \\
&= \quad \sum_j p(\theta_t = i|\theta_{t+1} = j, y_1, \ldots, y_t) \\
&\quad \times \quad p(\theta_{t+1} = j|y_1, \ldots, y_T)
\end{aligned}
$$

since θ_t is independent of y_s for $s > t$ given θ_{t+1}. Thus by an application of Bayes theorem we find

$$
\begin{aligned}
p(\theta_t = i|y_1, \ldots, y_T) &= \sum_j \frac{p(\theta_{t+1} = j|\theta_t = i, y_1, \ldots, y_t)p(\theta_t = i|y_1, \ldots, y_t)}{p(\theta_{t+1} = j|y_1, \ldots, y_t)} \\
&\quad \times\ p(\theta_{t+1} = j|y_1, \ldots, y_T) \\
&= \sum_j \frac{p(\theta_{t+1} = j|\theta_t = i)p(\theta_t = i|y_1, \ldots, y_t)}{p(\theta_{t+1} = j|y_1, \ldots, y_t)} \\
&\quad \times\ p(\theta_{t+1} = j|y_1, \ldots, y_T),
\end{aligned}
$$

where the last equality is due to the fact that θ_{t+1} is independent of $y_1, \ldots, y_t$ given θ_t. Next note that by an application of the law of total probability and the definition of conditional probability we have

$$
\begin{aligned}
p(\theta_{t+1} = j|y_1, \ldots, y_t) &= \sum_k p(\theta_{t+1} = j, \theta_t = k|y_1, \ldots, y_t) \\
&= \sum_k p(\theta_{t+1} = j|\theta_t = k)p(\theta_t = k|y_1, \ldots, y_t)
\end{aligned}
$$

so that all factors necessary for obtaining the marginal posterior distribution at location t given the data for the entire sequence is available after finishing the forward recursion. Hence we have a recursion for $p(\theta_t|y_1, \ldots, y_T)$ that runs in reverse direction and involves $p(\theta_t|y_1, \ldots, y_t)$ which we calculated using the forward recursion (in addition to a distribution specified by the model, i.e. $p(\theta_{t+1}|\theta_t)$). This recursion is referred to as the *backward recursion* in the sequence alignment literature and is called *smoothing* in other applications of state space models. So we obtain the marginal posterior distribution of the state process at each location by running the forward recursion, saving these results, then running the backward recursion. Finally, note that when there are only a few states, these computations will have a computational burden that grows linearly with the size of T since all summations will be rapid.

The backward recursion for the binomial example

Again we return to the example first discussed in Section 9.2.1 but now to provide an example of the backward recursion. Similar to the use of the forward recursion, suppose we have already computed $p(\theta_{t+1} = 0|y_1, \ldots, y_T)$ and $p(\theta_{t+1} = 1|y_1, \ldots, y_T)$ and stored these values (again, here we really only need one of these).

Using the formula for the backward recursion we then find that

$$
\begin{aligned}
p(\theta_t = 1|y_1, \ldots, y_T) &= \sum_i \frac{p(\theta_{t+1} = i|\theta_t = 1)p(\theta_t = 1|y_1, \ldots, y_t)}{\sum_j p(\theta_{t+1} = i|\theta_t = j)p(\theta_t = j|y_1, \ldots, y_t)} \\
&\quad \times\ p(\theta_{t+1} = i|y_1, \ldots, y_T).
\end{aligned}
$$

Which, using the $p_{i,t}$ notation introduced above, gives

$$
\begin{aligned}
p(\theta_t = 1|y_1, \ldots, y_T) &= p_{1,t}\left[\frac{(1-\phi_4)p(\theta_{t+1} = 0|y_1, \ldots, y_T)}{(1-\phi_3)p_{0,t} + (1-\phi_4)p_{1,t}}\right. \\
&\quad + \left.\frac{\phi_4 p(\theta_{t+1} = 1|y_1, \ldots, y_T)}{\phi_3 p_{0,t} + \phi_4 p_{1,t}}\right].
\end{aligned}
$$

Hence starting this recursion with the result of the end of the forward recursion allows one to characterize the marginal posterior distribution of θ_t for all t given values for the parameter vector ϕ.

9.3.3 The posterior mode of the state sequence

Another property of the posterior distribution of interest is the posterior mode over the entire sample space of the state sequence. That is, we are interested in maximizing the function $p(\theta_1, \ldots, \theta_T | y_1, \ldots, y_T)$ with respect to the vector $(\theta_1, \ldots, \theta_T)$. This is not the same as finding the sequence of θ_t that maximizes the univariate functions $p(\theta_t | y_1, \ldots, y_T)$ since there will be potentially strong associations between the values of the set of θ_t. We will use dynamic programming to find this sequence. To this end we need a recursion for the maximum of the desired function.

Now

$$
\begin{aligned}
p(\theta_1, \ldots, \theta_t | y_1, \ldots, y_t) &\propto p(y_t | \theta_1, \ldots, \theta_t, y_1, \ldots, y_{t-1}) \\
&\quad \times p(\theta_1, \ldots, \theta_t | y_1, \ldots, y_{t-1}) \\
&\propto p(y_t | \theta_t)p(\theta_t | \theta_{t-1})p(\theta_1, \ldots, \theta_{t-1} | y_1, \ldots, y_{t-1})
\end{aligned}
$$

so that

$$
\begin{aligned}
\mathrm{argmax}_{\{\theta_1, \ldots, \theta_t\}} p(\theta_1, \ldots, \theta_t | y_1, \ldots, y_t) &= \mathrm{argmax}_{\{\theta_1, \ldots, \theta_t\}} p(y_t | \theta_t)p(\theta_t | \theta_{t-1}) \\
&\quad \times p(\theta_1, \ldots, \theta_{t-1} | y_1, \ldots, y_{t-1}).
\end{aligned}
$$

This provides a recursion for the maximum and the sequence of θ_t for $t = 1, \ldots, T$ that achieves this maximum. Hence we use dynamic programming to sequentially update the maxima using this recursion. To make the connection to traversing a lattice that was used to motivate dynamic programming, consider a lattice where the rows are the possible states and the columns are the locations. Our goal is to then find the path that goes through one state at each location. The scores for the line segments are determined by the values of $p(y_t | \theta_t)p(\theta_t | \theta_{t-1})$. For example, if there are 2 states for the hidden process, then at each location there are 2 sites where each site has 2 edges from the previous position that lead to it. Label the states 0 and 1 and consider the site for state 1 at location t. This would have an edge that leads from state 1, and this edge would have a multiplicative contribution to the score given by $p(y_t | \theta_t = 1)p(\theta_t = 1 | \theta_{t-1} = 1)$, while the edge that leads from state 0 would have a multiplicative contribution to the score of $p(y_t | \theta_t = 1)p(\theta_t = 1 | \theta_{t-1} = 0)$. This algorithm is called the *Viterbi algorithm*.

9.4 Parameter estimation

In the previous 3 sections, we treated the parameters appearing in the conditional distributions that specify our hidden Markov model as known. In practice these parameters must be estimated from the data. We will now show how to compute the likelihood of these parameters. If there is prior information about the parameters, we could incorporate such information in the usual fashion. A popular approach to parameter estimation is to use the EM algorithm to estimate the parameters in the model (treat the state of the process as missing data so that during the M-step we are simply conducting inference for a Markov chain), then use these estimates as if they were known. If there are many locations relative to the number of model parameters we can think of this approach as an empirical Bayes approach that is a reasonable approximation to the full Bayes approach described here. The most popular implementation of the EM algorithm for hidden Markov models is called the Baum Welch algorithm (Baum, 1972).

Now, by manipulating the definition of conditional probability

$$p(\phi|y_1, \ldots, y_T) = \frac{p(\theta_1, \ldots, \theta_T, \phi|y_1, \ldots, y_T)}{p(\theta_1, \ldots, \theta_T|y_1, \ldots, y_T, \phi)}.$$

Hence we will have an expression for the posterior distribution of the model parameters if we can compute each of the 2 factors on the right hand side. For the numerator, we use Bayes theorem and the conditional independence of the y_ts given the θ_ts

$$p(\theta_1, \ldots, \theta_T, \phi|y_1, \ldots, y_T) \propto p(y_1, \ldots, y_T|\theta_1, \ldots, \theta_T, \phi)p(\theta_1, \ldots, \theta_T, \phi)$$

$$\propto p(\theta_1, \ldots, \theta_T, \phi) \prod_{t=1}^{T} p(y_t|\theta_t, \phi)$$

$$\propto p(\phi)p(\theta_1|\phi)p(y_1|\theta_1, \phi) \prod_{t=2}^{T} \Big[p(y_t|\theta_t, \phi)p(\theta_t|\theta_{t-1}, \phi) \Big].$$

For the denominator, we use the Markov property and manipulations identical to those used in the derivation of the backward recursion

$$p(\theta_1, \ldots, \theta_T|y_1, \ldots, y_T, \phi) = p(\theta_T|y_1, \ldots, y_T, \phi) \prod_{t=1}^{T-1} p(\theta_t|\theta_{t+1}, y_1, \ldots, y_T, \phi)$$

$$= p(\theta_T|y_1, \ldots, y_T, \phi) \prod_{t=1}^{T-1} p(\theta_t|\theta_{t+1}, y_1, \ldots, y_t, \phi)$$

$$= p(\theta_T|y_1, \ldots, y_T, \phi) \prod_{t=1}^{T-1} \frac{p(\theta_{t+1}|\theta_t, \phi)p(\theta_t|y_1, \ldots, y_t, \phi)}{p(\theta_{t+1}|y_1, \ldots, y_t, \phi)}.$$

Therefore we can express the numerator and denominator in terms of the filtering densities, the transition probabilities between states and the probability distribution of the observations given the states. If we put the previous expressions together we find that

$$p(\phi|y_1,\ldots,y_T) \quad \propto \quad \frac{p(\phi)p(\theta_1|\phi)p(y_1|\theta_1,\phi)}{p(\theta_T|y_1,\ldots,y_T,\phi)} \prod_{t=2}^{T} \frac{p(y_t|\theta_t,\phi)p(\theta_t|y_1,\ldots,y_{t-1},\phi)}{p(\theta_{t-1}|y_1,\ldots,y_{t-1},\phi)}.$$

Thus, to evaluate the likelihood for ϕ at a given value of ϕ, we run the forward algorithm to get $p(\theta_t|y_1,\ldots,y_t,\phi)$ for all t, then use these along with the 2 conditional probabilities that specify the model in order to evaluate the likelihood. Note that the right hand side of the equation for the likelihood of ϕ depends on θ_t whereas the left hand side does not. This implies that the likelihood is given by this expression for any value of $\theta_1,\ldots,\theta_T$, hence for purposes of numerical stability, we evaluate the right hand side at the marginal modes argmax $p(\theta_t|y_1,\ldots,y_t)$ since these are easy to evaluate if we run the forward recursion as we evaluate the likelihood. We can then maximize this expression as a function of ϕ using any number of numerical maximization routines. This will give the posterior mode of ϕ. This approach was first described in Reilly, Gelman and Katz (2001).

Parameter estimation for the binomial example

Once again we consider the example first discussed in Section 9.2.1 but now show how the method we just described for finding MLEs or the posterior mode of the parameters in the vector ϕ. We will suppose that all 4 parameters have priors that are uniform on the interval $(0,1)$. In addition, we suppose that both states are equally likely for the first location, so that $p(\theta_1|\phi) = \frac{1}{2}^{\theta_1}\frac{1}{2}^{1-\theta_1}$. This is just $p(\theta_1|\phi) = \frac{1}{2}$ and so we can ignore this factor. The quickest way is to just use the sequence of θ_t which have the property that θ_t is the most likely value of θ_t given by the forward recursion. Suppose we have this, a sequence of zeros and ones, denoted $\hat{\theta}_t$ where $\hat{\theta}_t = \arg\max_i p_{i,t}$. We then evaluate the expression for the likelihood at $\hat{\theta}$. The factor $p(y_t|\hat{\theta}_t)$ will just be the likelihood for a binomial observation. The success probability will be ϕ_1 if $\hat{\theta} = 0$ otherwise the success probability will be ϕ_2 which we can write as

$$p(y_t|\hat{\theta}_t,\phi) \propto [\phi_2^{y_t}(1-\phi_2)^{n-y_t}]^{\hat{\theta}_t}[\phi_1^{y_t}(1-\phi_1)^{n-y_t}]^{1-\hat{\theta}_t}.$$

In addition we need an expression for $p(\hat{\theta}_t|y_1,\ldots,y_t,\phi)$, but this will be determined while running the forward recursion and is given by

$$p(\hat{\theta}_t|y_1,\ldots,y_t,\phi) = \max_i p_{i,t}.$$

Finally we need an expression for $p(\hat{\theta}_t|y_1,\ldots,y_{t-1},\phi)$. To this end use the fact that

$$p(\hat{\theta}_t|y_1,\ldots,y_{t-1},\phi) = p(\hat{\theta}_t|\theta_{t-1}=0)p_{0,t-1} + p(\hat{\theta}_t|\theta_{t-1}=1)p_{1,t-1}$$

and note that

$$p(\hat{\theta}_t|\theta_{t-1}) = [\phi_4^{\hat{\theta}_t}(1-\phi_4)^{1-\hat{\theta}_t}]^{\theta_{t-1}}[\phi_3^{\hat{\theta}_t}(1-\phi_3)^{1-\hat{\theta}_t}]^{1-\theta_{t-1}},$$

so that

$$p(\hat{\theta}_t|y_1,\ldots,y_{t-1},\phi) = [\phi_3^{\hat{\theta}_t}(1-\phi_3)^{1-\hat{\theta}_t}]p_{0,t-1} + [\phi_4^{\hat{\theta}_t}(1-\phi_4)^{1-\hat{\theta}_t}]p_{1,t-1}.$$

Hence we can put all these expressions together to obtain

$$p(\phi|y_1,\ldots,y_T) \propto \frac{\left[\phi_2^{y_1}(1-\phi_2)^{n-y_1}\right]^{\hat{\theta}_1}\left[\phi_1^{y_1}(1-\phi_1)^{n-y_1}\right]^{1-\hat{\theta}_1}}{\max_i p_{i,T}}$$

$$\times \prod_{t=2}^{T} \frac{\left[\phi_2^{y_t}(1-\phi_2)^{n-y_t}\right]^{\hat{\theta}_t}\left[\phi_1^{y_t}(1-\phi_1)^{n-y_t}\right]^{1-\hat{\theta}_t}}{\max_i p_{i,t-1}}$$

$$\times \left\{\left[\phi_3^{\hat{\theta}_t}(1-\phi_3)^{1-\hat{\theta}_t}\right]p_{0,t-1} + \left[\phi_4^{\hat{\theta}_t}(1-\phi_4)^{1-\hat{\theta}_t}\right]p_{1,t-1}\right\}.$$

We could then obtain the posterior mode or MLE by maximizing this expression with some nonlinear optimization method. In practice one should maximize the log of this expression for numerical stability. For example, we have used the function nlminb as implemented in the software S-PLUS and found that the algorithm finds the posterior mode in seconds (with $T = 200$). One could potentially develop a faster algorithm by just setting $\hat{\theta} = 0$ (since then the forward recursion would not be needed to evaluate the likelihood), but in experiments the function nlminb will not converge even when the starting value is the true parameter value. Finally, note that in order to evaluate the posterior of ϕ one only needs the current value from the forward recursion and the values from the previous location, hence one should evaluate the likelihood and perform the forward recursion at the same time rather than first perform the forward recursion and save these results, then compute the likelihood.

9.5 Bayesian estimation of the state integrating out the model parameters

In practice, the distribution of the state sequence is of primary importance. Previously we treated ϕ as known, but now we have an expression for the posterior distribution of this parameter. If the sequence is long (i.e. T is large) and there are not many parameters (i.e. ϕ is of low dimension) then treating the posterior mode (or MLE) of ϕ as the known value of ϕ is often a useful approximation (we can interpret this as an empirical Bayes approach as usual). On the other hand, if there are many parameters and not much data, then ignoring uncertainty in the estimation of ϕ can lead to dramatic underestimation of the uncertainty in the estimation of the state process $\{\theta_t\}_{t=1}^{T}$. In this case we need to consider the uncertainty in the estimation of ϕ.

Although a number of approaches are possible, the following is typically easy to implement and works well compared to the popular alternatives. We will use a Bayesian simulation based approach. Our goal will be to obtain samples from the posterior distribution of the state process. We can then use these samples to characterize any aspect of the posterior distribution in which we are interested. For example, if we are interested in the posterior mean of θ_t for some t we just find the mean of our samples for θ_t. With enough samples, we can make this approximation to the mean as accurate as we desire.

Now

$$p(\theta_1, \ldots, \theta_T | y_1, \ldots, y_T) = \int p(\theta_1, \ldots, \theta_T, \phi | y_1, \ldots, y_T) \, d\phi$$
$$= \int p(\theta_1, \ldots, \theta_T | \phi, y_1, \ldots, y_T) p(\phi | y_1, \ldots, y_T) \, d\phi.$$

We can use this last equation as a basis for our simulations since we can evaluate the integral using Monte Carlo integration. That is, since we have an expression for the posterior distribution of ϕ we can draw samples from this distribution. Given a sample from the posterior of ϕ, we then simulate from $p(\theta_1, \ldots, \theta_T | \phi, y_1, \ldots, y_T)$ by using the recursion

$$p(\theta_1, \ldots, \theta_T | \phi, y_1, \ldots, y_T) = p(\theta_T | \phi, y_1, \ldots, y_T) \prod_{t=2}^{T} p(\theta_{t-1} | \theta_t, \phi, y_1, \ldots, y_T).$$

So we draw samples from each of the conditional distributions in turn conditional on the draw from the following value in the sequence. In this manner, the proper amount of association is induced in the sequence of θ_t for $t = 1, \ldots, T$. To draw a sample from the distribution $p(\theta_{t-1} | \theta_t, \phi, y_1, \ldots, y_T)$, we use the previously exploited relationship

$$p(\theta_{t-1} | \theta_t, \phi, y_1, \ldots, y_T) \quad \propto \quad p(\theta_t | \theta_{t-1}, \phi) p(\theta_{t-1} | y_1, \ldots, y_{t-1}, \phi). \quad (9.3)$$

We showed when deriving the backward recursion in Section 9.3. To use this expression to draw samples from the posterior, we draw the final state from its posterior, then draw samples from the other states conditional on the draw of the previous state. Since our state process has only finitely many states, we are just drawing from a multinomial distribution at each location.

Thus the steps are

1. simulate many draws from the posterior distribution of ϕ
2. for each simulated ϕ run the forward recursion to obtain the necessary conditionals $p(\theta_t | y_1, \ldots, y_t, \phi)$
3. simulate the state process in the reverse direction using equation 9.3.

The last 2 steps in this algorithm are referred to as the forward filtering, backward sampling algorithm (see the references Carter and Kohn, 1994 and Frühwirth-Schnatter, 1994). One can then calculate any aspect of the posterior distribution that is necessary for the scientific objective. For example, suppose one was interested in the first time (reading locations from left to right, say) that a state was reached with 95% certainty. This is easy to evaluate if we have many samples from the joint posterior distribution of all the states since one can just compute the frequency with which the event occurs in the simulations.

9.5.1 Simulating from the posterior of ϕ

While we have shown how to evaluate the marginal posterior distribution of the model parameters, ϕ, we did not describe how to draw simulations from this distribution. Since one can evaluate the distribution, one can draw a graph of the marginal distributions of each element of the vector ϕ. If these are approximately normal, as we expect on theoretical grounds, then we can approximate the posterior with a multivariate normal distribution (getting the covariance matrix from the negative inverse of the matrix of second derivatives of the log posterior distribution). Alternatively, if there are only 2 or 3 elements in the vector ϕ, then one can calculate the posterior over a grid of values in the parameter space and sample from these conditionally, i.e. use $p(\phi) = p(\phi_1|\phi_2)p(\phi_2)$ where subscripts index elements of the vector ϕ.

If the posterior is not well approximated as a normal distribution and the dimension of ϕ is too large to sample over a grid, we can use the Metropolis algorithm presented in section 2.6. For parameters that are constrained, such as success probabilities, we would want to sample on an unbounded scale since the Metropolis algorithm as described in section 2.6 works better in that case. For example, if $\phi_1 \in [0, 1]$, we would set $\eta_1 = \log \frac{\phi_1}{1-\phi_1}$ and sample from the posterior distribution of η_1 using a normal proposal distribution, then convert back to ϕ_1 to allow interpretation of the results.

9.5.2 Using the Gibbs sampler to obtain simulations from the joint posterior

Perhaps the most common method for obtaining samples from the posterior distribution of the states and model parameters is to use the Gibbs sampler. For hidden Markov models, one would sample the state process given the model parameters as we have outlined above, then simulate the model parameters conditional on the values of the state. It is simple to simulate from the posterior distribution of the model parameters given the values of the state process because this is identical to simulating from the posterior distribution for fully observed Markov models.

The drawback of using the Gibbs sampler comes from the fact that we are using the Metropolis algorithm over a much larger sample space compared to the approach advocated above. In practice, it is difficult to demonstrate that a chain has converged to its limiting distribution, and this problem becomes more difficult as the number of parameters increases and as the correlation between parameters in the posterior distribution increases. For hidden Markov models, typically T is large (e.g. 1000) and the number of model parameters is small (e.g. 20). Hence the Gibbs sampler constructs a Markov chain over a very large parameter space (e.g. 1020) compared to the method that first simulates the model parameters then simulates the state without a Markov chain (e.g. 20). In addition, the correlation between adjacent states in the posterior distribution is typically quite high (since they represent values of a Markov chain).

9.6 Exercises

1. Consider a 2 state Markov chain with transition probabilities p_1 and p_2. Find the limiting distribution of the chain.

2. Consider a 2 state Markov chain with transition probabilities p_1 and p_2. Explicitly find the maximum likelihood estimates of p_1 and p_2 given a sequence of observations $y_1, \ldots, y_T$ from a Markov chain. Give a condition on the identifiability of these estimates.

3. Suppose we observe a multinomial random variable, y_t, as we traverse a DNA molecule in a fixed direction that has one of 2 sets of probabilities, ϕ_1 and ϕ_2, and the probability of success is determined by a 2 state hidden Markov chain with transition probabilities ϕ_3 and ϕ_4. Give explicit forms for the forward and backward recursions.

4. Suppose we observe a multinomial random variable, y_t, as we traverse a DNA molecule in a fixed direction that has one of 2 sets of probabilities, ϕ_1 and ϕ_2, and the probability of success is determined by a 2 state hidden Markov chain with transition probabilities ϕ_3 and ϕ_4. Give details for use of the EM algorithm for estimation of the parameters ϕ_i by treating the value of the state process as missing data and optimizing with respect to ϕ_i.

5. Write a program to simulate a hidden Markov model with 200 locations supposing that there are 2 hidden states and y_t is multinomial with 4 categories. Suppose the probabilities for the categories are given by $(0.15, 0.05, 0.3, 0.5)$ in one state and $(0.5, 0.3, 0.05, 0.15)$ while the chain is in the other state. In addition assume that the chain stays in its current state with probability 0.9 for both states. Implement the forward and backward recursions and compare your solutions to the underlying hidden process.

6. Write a program to simulate a hidden Markov model with 200 locations supposing that there are 2 hidden states and y_t is multinomial with 4 categories. Suppose the probabilities for the categories are given by $(0.15, 0.05, 0.3, 0.5)$ in one state and $(0.5, 0.3, 0.05, 0.15)$ while the chain is in the other state. Find the posterior mode of the model parameters and show that your solution is close to the values used to generate the data.

CHAPTER 10

Feature Recognition in Biopolymers

A number of problems in computational biology can be cast as attempting to identify instigations of some known sequence feature. As the most simple example, we may seek to find locations of start codons in a sequence of DNA. Generally when given just a sequenced small part of a chromosome, one doesn't know the reading frame or on which strand to expect the start codon. Recall, the reading frame of a DNA sequence indicates which nucleotide is the first nucleotide of a codon, hence just given a sequence of DNA there are 3 possible reading frames for each of the 2 strands. Recall that which end of a DNA sequence is the 3' end and which is the 5' end is known from the sequencing reaction since the sequence is determined by termination of a growing chain of nucleotides (see section 3.1). Hence if we are given the following sequence of nucleotides from a sequencing experiment

AACAAGCGAA TAGTTTGTT

we have the following 3 sets of codons (and partial codons) that are possible on this strand

AAC AAG CGA ATA GTT TTG TT
A ACA AGC GAA TAG TTT TGT T
AA CAA GCG AAT AGT TTT GTT

whereas the other strand would have the following nucleotides (recalling that the 3'-5' orientation of the other strand runs in the reverse direction)

AACAAAACTA TTCGCTTGTT

we get the following set of possible codons

AAC AAA ACT ATT CGC TTG TT
A ACA AAA CTA TTC GCT TGT T
AA CAA AAC TAT TCG CTT GTT.

For this simple problem we could just scan each of these sets of codons and look for a start codon. Unfortunately, most sequence features of interest are far more complex in that certain locations in the common subsequence needn't be conserved across different instigations (in the same genome or across genomes from different species). For example, there are 3 distinct stop codons. We use the term *regulatory element* as a general term to describe subsequences whose presence in a sequence has some functional consequence on the use of that sequence. As an example of a regulatory element, a common subsequence that is identified by certain proteins that attach to DNA molecules called a spaced dyad motif can be described as 2 short (i.e. 3-5 bp) highly conserved (i.e. nearly identical in different organisms) sequence separated by a gap of highly variable sequence (and of variable length). A common interpretation for the existence of such subsequences is that a protein complex composed of 2 subunits binds to the DNA molecule at 2 locations but does not contact the DNA molecule for locations between these 2 subregions. Other examples exist where dependence over a number of bp in a subsequence is the critical feature. For example, a subsequence may behave as a regulatory element if there is a G in position 1 and a T or a A in position 6 of a subsequence. In reality the operation of such regulatory elements is likely extremely complex and should be understood as competing against other regulatory elements nearby in the same sequence for binding to nearby proteins.

Similarly, proteins act by binding to other molecules, hence we would like to determine what aspect of a protein sequence is necessary for binding to other molecules. One is often also interested in regulatory elements for RNA. For example, the process of RNA degradation is mediated by proteins that bind to the RNA at certain locations. If we know about these binding sites we may be able to predict decay rates based on sequence information. While these are important problems, we will here focus on regulatory elements in DNA. Studying DNA is simpler than RNA or proteins because DNA has no secondary structure, and the presense of secondary structure makes the discovery of binding sites much harder. On the other hand, the manner in which DNA is packed into chromatin must have profound implications for genetic regulation, but the implications of chromatin structure on genetic regulation are poorly understood currently.

Here we will consider 2 widely studied problems in this context: identification of transcription factor binding sites in DNA sequences and identification of genes in genomic DNA. Methods for gene detection frequently look for regulatory elements that govern the splicing of pre-mRNA. These problems have a number of features in common. First, both suppose something is known about the features we seek in terms of their size and the probability of the occurrence of the various nucleotides. The goal is to then use this information to predict the occurrence of transcription factor binding sites or aspects of a gene (e.g. transcription start sites and endpoints of exons). In addition, knowledge of transcription factor binding sites assists in the prediction of gene locations since they bind upstream of the gene they regulate (although quite far upstream in some cases). Finally, a number of the techniques employed in both analyses are quiet similar, as we shall see. The term *cis* is often used in biology and chemistry to refer to something being on the same side, and this term is used to refer to regulatory elements of a sequence that are in that sequence. On the other hand, transcription factors are *trans*-acting since these are typically proteins encoded in other regions of the genome. Prior to the treatment of these problems, we first discuss gene transcription in eukaryotes in greater detail to further understand the role of transcription factor binding as it relates to the control of gene expression.

10.1 Gene transcription

The process of gene expression is regulated by many different proteins. In order for a gene to be transcribed, dozens of proteins must bind to the DNA molecule in a complicated complex of proteins. These protein complexes bind to DNA in certain areas that are called transcription control regions (or *regulatory regions*). These complexes either promote or inhibit transcription (or work with other proteins whose function is to regulate transcription) and are referred to as transcription factors. How the regulatory elements that transcription factors bind to are organized on the genome has implications for gene expression and can help us locate genes in DNA. In addition, understanding regulation of gene expression is important for understanding disease processes. Currently, in association studies using SNPs, the conventional approach is to genotype some SNPs at the intron and exon boundaries and SNPs near the start of candidate genes in addition to SNPs in the genes themselves. This way, we can try to determine if disease states are associated with differences in gene regulation as opposed to differences in alleles for some candidate gene (i.e. a gene we think is linked to a disease locus due to prior knowledge).

The structure of the genome differs markedly between prokaryotes (organisms with cells that don't have nuclei) and eukaryotes. In the former case, the genome is organized into functional units known as *operons*. An operon is a segment of DNA that encodes a set of genes necessary for some function. Directly upstream from the operon is the transcription control region. In addition, some transcription factors in prokaryotes regulate the transcription of multiple operons: such sets of operons are referred to as regulons. Eukaryotes do not have operons and the regulatory elements are spread over the genome. The sites that regulate a gene in a eukaryote may be 1000s of basepairs from the gene. Hence the discovery of regulatory elements in eukaryotic DNA is fundamentally different, and harder, than finding such elements in prokaryotes. Here we focus on the eukaryotic case.

The catalyst for transcription of all eukaryotic protein-coding genes is a molecule known as RNA polymerase II (called Pol II). There are several known regulatory elements for this molecule. First, there is an element known as the TATA box that is often (but not always) found about 25-40 basepairs upstream from the start of a gene. In addition, there is a CA dinucleotide that is found very near, or right at the start, of a transcription site (this promoter element is called an *initiator*). Some genes do not have an initiator or a TATA box. Such genes have variable start sites, so that the same gene can give rise to differing mRNAs. Most of these genes have a region that has a high GC content of length about 20-50 bp about 100 bp upstream from the transcription start site. These control elements are referred to as *CpG islands*. In addition to these promoters, there are regions known as *enhancers* that are far from the start site and control promoter regions. Some enhancers are up to 50,000 bp from the promoter they regulate. Some enhancers can bind multiple transcription factors. In addition, thousands of individual binding sites have been recognized that are capable of binding hundreds of transcription factors. Many mammalian genes have multiple enhancer regions. As the previous examples illustrate, most regulatory elements in DNA are very short, and hence difficult to detect. Given the difficulty of locating short segments in a sequence of DNA, three basic strategies have been used to find regulatory elements: look for groups of elements in a sequence of DNA, look for the same element in homologous genes in related organisms (i.e., *phylogenetic footprinting*, see Wasserman et al., 2000), or look for regulatory elements for co-expressed genes. If we have the protein sequence from a set of genes from closely related organisms or that are co-expressed, we can use the motif discovery and multiple alignment tools from Chapter 11. Here we will concentrate on methods that look for previously identified regulatory elements or groups of them.

10.2 Detection of transcription factor binding sites

A number of approaches have been developed for identification of transcription factors. All require sets of experimentally verified elements and the goal is to detect new instigations of these elements in other sequences. There are a number of experimental techniques for identification of transcription factor binding sites. The most basic approach is based on the discovery that regions of the genome that are said to be hypersensitive to the DNase I enzyme are regions where the chromatin is open. Such regions are likely to contain transcription factor binding sites since nuclear proteins have physical access to DNA. Further experiments can then be conducted on these hypersensitive regions. A shortcoming of these approaches is that they only indicate regions to which transcription factors bind without indicating what transcription factor binds to that site. In addition to the computational techniques discussed here, there are several high-throughput methods that can be used, and these indicate the transcription factor and the locations. These involve designing an antibody for a transcription factor, labeling this antibody (or fix it on a solid surface with known location), then allowing the labeled antibody to bind to the transcription factor while the latter is bound to a segment of a DNA molecule. One then separates the protein DNA mixture based on binding to the antibody. Following this, one determines the sequences of the DNA sequences either using a microarray (a technology discussed in the last 4 chapters) or sequencing (the former is called ChIP-chip and the latter is called ChIP-seq). There are 2 large databases of experimentally confirmed transcription factor binding sites: TRANSFAC (Matys et al., 2006) and JASPAR (Sandelin et al., 2004). These store the binding sites in the form of position specific scoring matrices (PSSMs), which are also referred to as positional weight matrices, and these will be discussed below.

10.2.1 Consensus sequence methods

In consensus sequence methods, one defines the consensus sequence of a set of binding sites and then searches through the sequence to find instigations of these consensus sequences. For example, if one had a set of aligned regulatory elements, one could determine which positions have complete agreement among all aligned sequences (some positions will just be variable). There are a number of methods for obtaining a multiple alignment of a set of nucleotide sequences, some of which are discussed in detail in Chapter 11. More sophisticated applications (e.g., Prestidge, 1995) look for several of these consensus sequences simultaneously and find regulatory regions based on the density of regulatory elements. Choices regarding how dense regulatory elements need to be are determined by comparing known regulatory sequences to sequences which are known to not be regulatory regions.

10.2.2 Position specific scoring matrices

The idea behind PSSMs is that if we have a set of binding sites, we can calculate a frequency distribution for the nucleotides at each position and use this information to scan a segment of DNA for possible binding sites. This is superior to the consensus sequence based methods in that we don't need a regulatory element to completely match the consensus sequence. Note that if we don't require consensus at a position in the context of the consensus based approach it is possible for none of the regulatory regions used to find the consensus sequence to have the consensus sequence in them, and this points to a potentially serious shortcoming of that approach. As noted above, this method is likely to produce many false positives if the binding sites are of short length since by chance many short subsequences will be necessarily similar in terms of any measure of similarity.

As a concrete example, we will consider the method of Quandt et al., (1995). This method uses the idea of entropy, which is a commonly used measure of the extent of disorder in a system. If X is a random variable so that $P(X = x_k) = p_k$, then the entropy associated with X is defined to be

$$H(X) = -\sum_k p_k \log p_k.$$

Note that if $p_k \to 1$ for some k then $H(X) \to 0$, otherwise $H(X)$ is a positive constant. In fact, $H(X)$ will be maximized when all the p_k are equal (see exercise 1), so that large $H(X)$ indicates that we know as little as possible about the outcome of one trial, whereas small values indicate we know with near certainty the outcome of a single trial. The first step is to construct a matrix, A, of frequencies for the nucleotides in each position. Denote the i, j entry of the matrix A with A_{ij}. If we have a binding site of length L, then this matrix is 5 by L if our binding sites have gaps (which we assume here), otherwise this matrix is 4 by L.

Then one assigns a score to each column of this matrix using a linear transformation of the estimated entropy for that column so that the scores lie between 0 and 100, with the latter indicating near complete agreement between all sequences used to construct the matrix at that site. If we use the notation $p(i, b)$ to represent our estimate of the probability of observing base (or gap) b at position i in the binding site, then the associated estimate of the entropy is just $-\sum_b p(i, b) \log p(i, b)$. As noted above, if this is near zero then there is a high degree of conservation at that position in the binding site. We need to qualify this statement with "near" since the usual unbiased estimate of a proportion (i.e. the sample proportion) would be zero if there was complete agreement and $\log(0)$ is undefined (the authors suggest the use of the convention, $0 \log 0 = 0$ to avoid this problem). As an alternative, one could use a slightly biased estimated of the proportion that adds a 1 to each element of A so that no sample proportions are exactly zero after adding this value (this can be interpreted as the use of a certain prior, as discussed in exercise 2). In either event, since the entropy is maximized when $p_k = \frac{1}{5}$, we have that

$$0 \le H(X) \le \log(5)$$

so that

$$0 \ge -H(X) \ge -\log(5)$$

which implies

$$\log(5) \ge -H(X) - \log(5) \ge 0$$

so that if we use the scores

$$c_i = (100/\log 5)\left[\sum_b p(i,b) \log p(i,b) + \log 5\right],$$

we have a set of scores (one for each position in our binding site) that are between 0 and 100 with 100 indicating a higher degree of conservation at that position in the binding site.

Given the set of c_i, the score of a query sequence, x, with element given by the numbers $1, \ldots, 4$ is given by

$$\frac{\sum_i A_{x_i,i} c_i}{\sum_i \mathrm{argmax}_j A_{j,i} c_i}.$$

One then sets a threshold and searches through the query sequence to try to find subsequences that give a score higher than the threshold. Note that if there is 100% agreement at each site then $c_i = 100$ for all i. Also note that in this case, if one were to scan an exact match using this scoring scheme, then our score would be 1. For this reason the score reflects how much agreement exists between the binding site and a site under investigation. Moreover, the method accounts for not only the extent of a match, but also the level of entropy (i.e. uncertainty regarding one realization of a random deviate) at all locations in the binding site. The authors suggest that 0.85 is a reasonable cutoff value that is sufficiently high to find functional similarities for the examples they consider. Others (such as Levy and Hannenhalli (2002) have used the extreme value distribution to determine p-values for matches. Here the extreme value distribution is used because the highest scoring segment can be modeled as the maximum of a set of iid random variables (even though many of these segments overlap and so can't be independent, the effect is not so large for the short length of a typical regulatory element).

Improvements to these methods have been developed by use of more sophisticated models for the joint distribution of nucleotides within the regulatory region. For example, Hannenhalli and Wang (2005) used a mixture of PSSMs in place of a single PSSM and found improved prediction of binding sites. Another extension has been to model the dependence among the different positions rather than assume that these positions are independent. Osada, Zaslavsky and Singh (2004) found that modeling the dependence improved prediction of binding sites in *E. coli*.

10.2.3 Hidden Markov models for feature recognition

A number of approaches to the use of HMMs have been proposed in the literature. The basic idea behind all of these methods is that there is a hidden process that determines if the chain is in a transcription factor binding region or not. Here we describe some of these methods in greater detail.

A hidden Markov model for intervals of the genome

A simple HMM was proposed by Crowley, Roeder and Bina (1997). The basic idea is that there are regulatory regions that contain many regulatory elements, hence we search for these regions based on the presence of known regulatory elements. Although the authors do not do so, one could then use these identified regulatory regions to look for more regulatory elements. The first step is to construct a database of known (i.e. experimentally verified) regulatory elements. Let Δ denote the length of the longest regulatory element in our database. Then partition the sequence into I intervals of width Δ. Let X_{ij} be a variable that is 1 if regulatory element j is interval i. The hidden process, θ_i, is a binary process that is 1 if interval i is part of a regulatory region and 0 otherwise. We suppose that there is probability λ of entering a regulatory region if one is not in a regulatory region and there is a probability of τ of staying in a regulatory region. Note that θ_i is a Markov chain in this formulation. The authors use prior distributions on these 2 parameters so that we expect 0-4 regulatory regions every 5000 bp.

One problem with this model is that if one is in a regulatory region, then it is possible that one could observe a regulatory element from the assembled database just by chance, i.e. even if there is a regulatory element, it may not behave like a regulatory element. To bring this idea into the model, the authors introduce another unobservable variable, η_j that is 1 if regulatory element j is really a regulatory element for the sequence under investigation. They use the prior distribution that these η_j parameters are iid Bernoulli with parameter $\frac{1}{2}$. The distribution of X_{ij} is then specified as

$$X_{ij}|\theta_i = s, \eta_j = t \sim \text{Ber}\left(\phi_0 \frac{\sum_i X_{ij}}{\sum_{ij} X_{ij}}\right)$$

for $s = 0, 1$ and $t = 1$ or $s = 0$ and $t = 1$, and

$$X_{ij}|\theta_i = 1, \eta_j = 1 \sim \text{Ber}\left(\phi_1 \frac{\sum_i X_{ij}}{\sum_{ij} X_{ij}} \log\left(\frac{\sum_{ij} X_{ij}}{\sum_i X_{ij}}\right)\right).$$

Finally, they use a uniform prior on the set $\phi_0 < \phi_1$ since we expect to observe regulatory elements more frequently in regulatory regions.

While it may seem peculiar that the conditional distributions of the X_{ij} depend on a function of X_{ij} for all i, j, one can construct a model and that is equivalent to the previous model if one uses an empirical Bayes approach to estimate some of the model parameters. This is because, if we specified a model in terms of the conditional probability that the sequence contains the jth regulatory element given that the sequence contains some regulatory element, denoted π_j, and further supposed that the probability that the jth regulatory element acts as a regulatory element depends linearly on some function f of π_j, then we have the following model:

$$X_{ij}|\theta_i = s, \eta_j = t \sim \text{Ber}(\phi_0 \pi_j)$$

for $s = 0, 1$ and $t = 1$ or $s = 0$ and $t = 1$, and

$$X_{ij}|\theta_i = 1, \eta_j = 1 \sim \text{Ber}(\phi_1 f(\pi_j)).$$

The authors propose to use $f(x) = -x \log(x)$, so that the more one of the regulatory elements contributes to the entropy of the conditional distribution of observing one of the regulatory elements given that one is in some regulatory element, the more likely that one really is in some regulatory region. If one then uses the sample proportions to estimate the π_j, one obtains the model proposed by Crowley, Roeder and Bina.

A HMM for base-pair searches

Another HMM that has been used with success in locating regulatory elements looks for clusters of known regulatory elements (Frith, Hansen and Weng, 2001). For this method one also needs a set of known regulatory elements. The first step is the construction of a PSSM that gives the frequency of each nucleotide for each position of the binding site for each of the known regulatory elements. The model has a general background state, an intra-cluster background state and 2 sets of states for each regulatory element (one state for each DNA strand). The probability of a transition from a background state to any binding site is the same for all binding sites and strand orientations. Once one enters a binding site state, the chain moves through each bp of the binding site. The distribution for the observations given that one is in a binding site state is determined by the PSSM for that binding site and location in the binding site. For the background states, a moving average is used to determine the probability of observing each of the nucleotides. The framework is quite flexible and can be modified as more is learned about the nature of these binding sites. Moreover, one needn't break up the query sequence into distinct intervals as in Crowley, Roeder and Bina's approach.

10.3 Computational gene recognition

Estimating the presence of a gene in a DNA sequence is a problem that has attracted a great deal of attention and consequently there are many methods and software implementations available. A number of papers have compared these approaches, such as

Burset and Guigó (1996) and Guigó et al. (2000), and reveal that despite all this activity the accuracy of these methods is not completely satisfactory. There are several possible approaches to detecting genes. For instance, we may be interested in determining simply if a nucleotide in a DNA sequence is part of a gene. In contrast, a more difficult problem is determining the location of an exon. Finally, we might attempt to determine the locations of all the exons that comprise the protein coding regions of a gene. The latter can be quite difficult when one considers that some genes have hundreds of exons (on average, there are 9 exons in a human gene, with the gene titin having 363 exons, which is currently the largest number of exons in a human gene). The accuracy of methods for each of these 3 problems problems has been reported to be roughly on the order of 80%, 45% and 20% respectively by Zhang (2002). While experimental approaches are available, there is still a need for computational methods since such methods can reduce the necessary amount of experimental work. While there are several experimental approaches to identifying genes in DNA, these require specification of a short sequence where one thinks the gene resides. For example, one approach inserts the sequence under investigation into an intron in an artificial minigene (i.e. a sequence that has a transcription start site and 2 exons separated by an intron where the sequence under investigation is inserted). One then detects if the artificial gene with the inserted sequence has nucleotides from the inserted sequence. While several papers have examined the performance of the algorithms that have been proposed, generally the best method in such comparisons depends on the data set used to conduct the test. While exons do, by definition, become part of the mRNA molecule, some exons reside in a location of the mRNA molecule that are not translated (called the untranslated region (UTR)), however almost all gene prediction methods attempt to discover coding sequences, although this often isn't explicitly stated. There is a UTR found at each end of a mRNA molecule, and these are distinguished from each other via reference to the 5' UTR and the 3' UTR.

Almost all methods for detecting exons are based on an experimentally well validated model for the operation of the set of proteins that conduct the editing that allows pre-mRNA to become mRNA (this complex of proteins is referred to as the *spliceosome*). In this model, proteins bound to the pre-mRNA molecule allow for proteins that are part of the spliceosome to recognize the exon via molecular bonds. Hence one can recognize exons by the presence of certain sequences that are present near the ends of the exon. Thus if one has an experimentally determined set of exons, one can attempt to detect exons in a different DNA sequence by determining if certain subsequences nearly common across all the experimentally determined sequences are found in the new sequence. Framing the problem in this fashion allows for the use of many existing algorithms that have been developed in the context of statistical pattern recognition.

Not all exons are recognized with the same sensitivity. In particular, the 3' exon is the easiest exon to detect for several reasons. First, in the sequencing reaction that leads to the determination of the sequence that is entered into a databank such as Genbank, the 5' end of the transcript can be cut off due to technical aspects of the sequencing reaction; however there is now a database with millions of entries of uniquely mapped 5' end sequences (see Wakaguri et al., 2008). This will clearly make any method that attempts to detect genes using a known set of gene sequences problematic as one's basis for estimating the 5' end will itself be incorrect. In addition, the 3' end exon is the largest of the exons on average. Several features are also known about this exon, such as a high content of G and U nucleotides and the presence of certain short sequences comprised of A and U nucleotides (Graber et al., 1999). A number of algorithms and software implementations are available for detection of the 5' end of a transcript. These methods look for CpG islands and the TATA box in order to detect the transcription start site. Perhaps surprisingly, it is difficult to detect genes that don't have any introns. This is due in part to the presence of *pseudogenes* that have arisen due to 1 or more mutations in an ancestral gene that made that protein lose its function. In addition, most programs have difficulty determining if a sequence is a long internal coding exon or an intronless gene. The problem is magnified by the fact that few genes are intronless in humans, hence information in databases regarding such genes is naturally limited.

10.3.1 Use of weight matrices

Early methods used weight matrix approaches similar to the use of PSSMs in the context of transcription factor binding site recognition. Extensions of these approaches also allowed Markov dependence in the binding site so that complex PSSMs, in which the probability of observing certain nucleotides at a given location depend on the previous nucleotide, are used for gene detection. In typical applications one would develop a PSSM for an area near the start site of a gene or bridging the junction of an exon intron boundary. While other methods have proved superior, the use of weight matrices is still an important component of gene detection algorithms.

10.3.2 Classification based approaches

If one has data on subsequences that are typical of the start location of a gene or an exon intron boundary, one can use a variety of standard approaches to classification. A number of standard algorithms for classification will be discussed in Chapter 15, but the basic idea is to develop a decision rule to determine if a given sequence falls into 1 of 2 or more groups (e.g. is the subsequence part of the start site of a gene) based on data from sequences whose class we know. Approaches falling in these categories used tools such as also decision tress and neural networks (i.e. the software program GRAIL). However these methods don't give much improvement over weight matrix approaches. Substantial improvements require the use of information on multiple features simultaneously, for example, a splice site score, an exon score and an intron score can be combined using linear discriminant analysis (as in Solovyev, Salamov and Lawrence, 1994). One could also use quadratic discriminant analysis, and one implementation of this idea, the MZEF program (developed by Zhang, 1997), appears to perform better than either linear discriminant analysis or the neural network based approach. Nonetheless, in systematic comparisons of methods that use a single sequence feature to detect an exon, only about 50% of the exons known to be present are actually found (Burset and Guigó, 1996). Further improvements are possible by using information from multiple exons from a gene to detect all of them simultaneously.

10.3.3 Hidden Markov model based approaches

As noted above, since there are subtle relationships between the various exons in a gene, methods which incorporate information from multiply detected exons can improve estimation for each of the exons. For example, if the data indicate that one part of a sequence is a 5' exon and there appears to be evidence for 7 internal exons then a 3' exon, one would be more inclined to believe these internal exons are indeed exons than if one ignores the information on the 3' and 5' exons. In fact, such models can be improved using known sequence features in sequence that is adjacent to an exon. For example, we know that the 5' exon must start with the codon corresponding to M, so that the first 3 nucleotides are ATG. In addition, internal exons are known to have an AG pair just upstream and a GT downstream while 3' exons have an AG pair just upstream and end with a stop codon. From this perspective an intronless gene would begin with the start codon and end with a stop codon.

The simplest HMM for exon detection would introduce states corresponding to the start of the exon, a state for inside the exon and a state for the end of the exon. For each state there would be different probability distributions for observing nucleotides. The simplest example would be that nucleotides are iid with probabilities of observing different nucleotides depending on the state of the HMM. More complex approaches introduce mechanisms for inducing dependence among nucleotides in nearby locations, such as Markov dependence. Even more sophisticated models introduce further states so that multiple exons can be detected. For example, some methods (such as the method implemented in the program HMMGENE) look for promoter regions and regions that are used to attach the poly-A tail to the mRNA (called polyadenylation signals). By introducing a state corresponding to an intron one can also allow for multiple exons to be present in a sequence. One can use HMMs to find the most likely locations of coding regions by introducing states that correspond to the presence of these sequence features and then use standard tools from HMMs to estimate the state sequence.

More sophisticated algorithms have been developed that use a generalized HMM (GHMM), such as GENIE and GENSCAN (see Kulp et al., 1996). These have several features over the usual use of HMMs including more realistic distributions for the length of an intron and use of neural networks to detect splice locations. As shown in exercise 4, for a standard HMM, the distribution of the length of an exon will be geometric, however the observed distribution of exon lengths appears to be log normal. The use of semi-Markov models for the state sequence allows for more flexible exon lengths (such models are referred to as GHMMs). To formulate such a model, one again specifies a matrix of transition probabilities for the state sequence. The difference resides in the use of a length distribution that gives the number of nucleotide positions that the chain stays in the current state, $f_i(\eta)$, where i subscripts the set of possible states and $f_i(\eta)$ is a probability mass function on η, an integer. Estimates of the length distributions are obtained from smoothing the observed histograms of length distributions after stratifying exons by initial, internal or terminal. Exons are typically from 50 bp to 300 bp whereas introns are sometimes thousands on bp in length. This is consistent with the model of the spliceosome described above since the exon needs to bind to the spliceosome, hence there are restrictions on the lengths of exons, whereas introns are not required to bind other proteins hence their size is not as important.

Note that we can still define a sequence θ_t that describes the movement of the process through the Markov chain, however the value of θ_t only changes at a set of times determined by the values that η takes. Then, given that the hidden process is in state i for η nucleotides one uses the idea behind PSSMs to specify the joint distribution of the sequence data. That is, one draws the observed sequence $y_1, \ldots, y_\eta$ assuming that the y_j are independent with joint distribution

$$\prod_{j=1}^{\eta} p_j(y_j | \theta_j = i).$$

More complex models are frequently used that allow for Markov dependence in the distribution of the y_i, namely

$$p(y_1 | \theta_1 = i) \prod_{j=1}^{\eta} p_j(y_j | y_{j-1}, \theta_j = i).$$

For polyadenylation signals and promoters, previously developed models are used that describe the sequence features that have been found. Some implementations of these approaches use separate inhomogeneous, 5th order Markov transition matrices for 6-mers that end at each of the 3 codon positions to model the sequence distribution when in a coding region. For noncoding regions use of a homogeneous 5th order Markov model has been proposed. Such models have many parameters, hence extensive data is necessary to fit these models.

Further extensions and modifications of these methods have been proposed. For example, in Burge and Karlin (1997), the authors allow for different gene density and structure in regions where the GC content differs since there tend to be more genes in G-C rich regions of the genome. In addition, this method allows for multiple genes, genes on either strands and even just part of a gene (which is quite possible if one is looking for a gene in a random fragment of genomic DNA). Finally, these authors have models for the interaction between the endpoints of adjacent exon endpoints. We call the end of an exon the acceptor site and the beginning of the next exon the donor splice site. The authors then allow for complex interactions between elements of the donor sites (however this wasn't necessary for acceptor sites). Applying the same rationale used to justify looking for multiple exons would suggest that one should look for multiple genes when trying to detect any individual gene, as in the approach just described. However, this has not been convincingly done as of yet due to the difficulty of detecting genes in large genomic DNA sequences.

10.3.4 Feature recognition via database sequence comparison

In contrast to attempting to model the process by which the existence of genes is detected from the presence of various sequence elements, one could simply compare a query sequence to sequences of known genes in a database. A number of programs have been developed that attempt to do this, such as PROCRUSTES. In Gish and States (1993) the authors first transcribe DNA into amino acid sequences since the similarity of 2 sequences is more easily detected at the protein levels (because we have substitution matrices for amino acids but not for DNA). This must be done for the 6 possible reading frames. Then database searches are conducted using BLAST (the software to do this is called BLASTX). In Gelfand, Mironov and Pevzner (1996) the authors first find a set of blocks that contain all of the possible exons by looking for the potential acceptor and donor sites for splicing (these are indicated by the AG and GU dinucleotides). One then attempts to assemble these blocks (realizing that many will be false positives) by assigning scores to various block assemblies that correspond to various exons structures. These structures are then assessed by comparing them to a database that contains protein sequences. Other implementations use sequence similarity and HMM based approaches, such as GENEWISE. While these methods can perform well, they are computationally intensive and on average seem to increase accuracy only slightly.

10.3.5 The use of orthologous sequences

A similar approach is to use sequence information from multiple species in order to detect regulatory regions nearly common across all of the species considered. In this approach one uses a generalized pair HMM to find orthologous base pairs under assumptions such as the same number of exons are in both sequences. For example, in Meyer and Durbin (2002) mouse and human DNA sequences are used to detect genes common to both. This approach has the potential to find new genes since we are not comparing to a database of known genes.

A pair HMM is used to describe the alignment and annotation (which are done simultaneously). For 2 sequences y and x with elements y_t and x_t, a pair HMM is specified by a set of transition probabilities for the state $p(\theta_t|\theta_{t-1})$ as before, but now we model the joint conditional probability $p(x_t, y_t|\theta_t)$. Then the Viterbi algorithm can be used to find the most likely state sequence. Since the storage requirements can become large when analyzing long sequences they propose to use the Hirschberg algorithm in place of Viterbi because it requires less storage (although it can be a little slower). The set of states is similar to the case where GHMMs are used, however here there is now the possibility that one of the sequences is in one state while the other sequence is in some other state. The program DOUBLESCAN implements a version of the method that includes a total of 54 states.

Similarly, Pedersen and Hein (2003) explicitly model the evolutionary process that relates the 2 sequences in addition to the use of HMMs as is common in the gene detection literature. McAuliffe, Pachter and Jordan (2004) use a similar approach although they emphasize that one should use more than 2 species, and they argue that methods that look for orthologous sequences work better if the species are more closely related than the species that are frequently used for this sort analysis. For example, they show that by using 5 primates they get excellent results. The problem with these approaches stem from the fact that many regions that are conserved across species are in non-coding regions, hence such methods will tend to produce too many false positives unless one introduces a state in the HMM that corresponds to the sequences being conserved but non-coding.

10.4 Exercises

1. Show that the entropy of a random variable X so that $P(X = x_k) = p_k$ is minimized when $p_k \to 0$. Show how to maximize the entropy.

2. Let X be a Bernoulli random variable with success probability θ. If x is the number of successes in a total of n trials, and we specify a beta prior with parameters a and b, for what values of a and b would

$$\hat{\theta} = \frac{x+1}{n+2}$$

be the posterior mean?

3. Find the MLE of the entropy associated with a discrete random variable that takes p values. Does this estimate always exist? Explain.

4. Derive the probability mass function for the length of time that a Markov chain stays in 1 state. How does this distribution compare to the log-normal distribution?

5. If $\hat{p}_i$ for $i = 1, \ldots, 3$ is the set of the frequencies for observing the 3 stop codons in humans, what is the PSSM for the stop codon?

6. Use the method of Quandt et al. (1995) described in section 10.2.2 to scan chromosome 2 (accession number NC_000002.10) for the PSSM for the human start site, given as (from Zhang (1998))

position	-9	-8	-7	-6	-5	-4	-3	-2	-1	0	+1	+2	+3
A	0.21	0.15	0.21	0.21	0.18	0.24	0.57	0.21	0.15	1.0	0.0	0.0	0.26
C	0.28	0.38	0.38	0.24	0.35	0.49	0.05	0.43	0.53	0.0	0.0	0.0	0.20
G	0.31	0.27	0.22	0.42	0.27	0.22	0.36	0.17	0.27	0.0	0.0	1.0	0.41
T	0.19	0.20	0.19	0.13	0.20	0.05	0.02	0.10	0.04	0.0	1.0	0.0	0.10

Are the potential start sites (using a cutoff of 0.85) uniformly distributed on the chromosome?

7. Use GENIE and GENSCAN to identify genes on chromosome 16 (accession number NC_000016.8). Do the results agree?

CHAPTER 11

Multiple Alignment and Sequence Feature Discovery

11.1 Introduction

While the need for comparing 2 sequences is clear, there are circumstances in which one wants to compare more than one sequence in order to construct the alignment. For example, we may have the sequence of amino acids from a set of instances of a particular protein from several organisms (i.e. a set of orthologuous sequences) and we would like to determine what subsequences of amino acids are common across all the organisms for which we have data (e.g. if we had the sequence of IL-7 from chimpanzees we would like to align the human, mouse and chimpanzee versions of the protein). The subsequences that are conserved across species are likely to be the parts of the protein that must be conserved in the face of evolutionary pressure for the protein to retain its function. Hence a multiple alignment can help identify such regions. In addition, the 3 dimensional shape of a folded protein is in part governed by the local interactions between the residues that make up the peptide sequence. Thus we expect that the presence of certain subsequences would impact the 3 dimensional structure of the protein, and the ability to discover such common subsequences may help towards the goal of computational prediction of protein structures.

Another context in which the problem of multiple alignment arises is when attempting to discover new regulatory regions in DNA sequences. In this application, we suppose that we have information about the nucleotides that are present in a part of a DNA sequence that controls the level of gene expression for a set of genes that all tend to increase or decrease their level of expression in response to changes in the cellular environment. In this application, the assumed reason for the co-regulation of these genes is that they all have some aspect of their DNA sequence that is common and is recognized by some, potentially unknown, transcription factor(s). In addition to the genes having some commonality in terms of their DNA sequences near the genes, this commonality may reside in an UTR of its mRNA molecule. As another example in this vein, one could have a set of proteins thought to be related from a number of species and one would like to see what aspects of these sequences (and their nearby promoter and enhancer regions) are conserved across all species (this sort of analysis is called phylogenetic footprinting). Note that this application could benefit from additional information regarding how related the species are, nonetheless many researchers have reported success using simple multiple alignment algorithms with a set of orthologuous sequences. Finally, a number of approaches have been developed that try to identify regulatory regions in sequences by combining information from genes from the same organism and the same set of genes from other organisms. As when aligning just 2 sequences, we can discuss global and local alignment methods, although for some of the methods described here, as currently used in practice, this distinction is not as clear as when aligning just 2 sequences.

A number of approaches to this problem have been developed. First, one can just attempt to extend the approaches used for aligning 2 sequences. Extensions of these methods are possible, and we will discuss dynamic programming and methods that are in some sense extensions of the rapid methods for sequence alignment discussed in Section 8.3. However, distinct approaches are possible since there are multiple sequences and these sequences themselves reveal not only how much information is shared between the sequences, but also how much we expect the sequences to differ. This allows the use of models that have parameters that can be estimated directly from the sequences themselves, in contrast to the situation where there are only 2 sequences and we need to investigate other sets of proteins to determine how much similarity to expect between the sequences if they really are functionally related.

11.2 Dynamic programming

One can extend the previous methods for aligning a pair of sequences using dynamic programming to allow one to align more than one sequence at a time, but the computational burden increases to the extent that this approach is not in widespread use. To understand the nature of the problem, suppose we are trying to align 3 sequences. For aligning 2 sequences, one constructed a 2 dimensional array (i.e. a matrix) and filled in this array with the scores determined in a fashion discussed earlier. When we compare 3 sequences, we must fill in a 3 dimensional array. Not only are there larger memory requirements, but the number of paths that lead to each internal element of the table goes from 3 (or 4 for local alignment) to 7 (3 ways to have a gap in 2 of the sequences, 3 ways to have a gap in 1 of the sequences and 1 way to have no gaps). This increases the computational burden to the extent that dynamic programming is no longer a practical method for finding the optimal alignment. Clearly one can consider local and global multiple alignment as in pairwise alignment.

There have been many attempts to speed up dynamic programming methods in the context of multiple alignment. The basic idea is that we shouldn't really calculate all entries in the dynamic programming array. Any alignment that deviates strongly from the main diagonal will entail many gaps, implying that the sequences shouldn't be aligned since they don't have much in common. The decision regarding which entries should be evaluated is what distinguishes these various methods. For example, one can look at all of the pairwise alignments. These give upper bounds on how well any 2 sequences can be aligned in a multiple alignment. One can use fast heuristic methods for multiple alignment (see Section 11.3 for an example) to get lower bounds on the multiple alignment, and these bounds reduce the number of entries in the dynamic programming array that must be evaluated.

11.3 Progressive alignment methods

There are a number of more heuristic methods for multiple alignment. Progressive alignment methods are based on the natural idea of examining all of the pairwise alignments and using these to construct a multiple alignment. There are many variations on this basic idea. What distinguishes most of these methods from the others is the manner in which one makes the choice regarding what pairwise alignments are most important for constructing the multiple alignment. The popular program CLUSTALW uses a progressive alignment method to construct multiple alignment. Although there are many details to this algorithm, the idea is as follows. One first calculates all pairwise alignment scores. Then one uses a cluster algorithm to determine what sequences are closest using the alignment scores. Cluster analysis is a set of techniques that will be discussed in further detail in Chapter 14, but the basic problem addressed is how to group observations into sets so that observations in the same set are similar and observations in different sets are dissimilar. In the next step one aligns the 2 closest sequences to get the pairwise alignment between them. This alignment is then modified by aligning the next closest sequence to the 2 used thus far to the alignment of the original 2 sequences. One then continues by aligning the next most closely related sequence to the alignment one has obtained thus far.

One can use online implementations of this program, or download the program and run it locally. Local download is necessary for certain computationally expensive analyses. To run the program one must specify a set of proteins in an appropriate format. One such format is the NBRF/PIR format that starts a new sequence with the > symbol, then gives a 2 character designator. The most common of these 2 character designators are P1 for complete proteins, F1 for protein fragments F1, DL for linear DNA and RL for linear RNA. Then one puts a semicolon followed by the data bank Identifier (which can be arbitrarily set by the user, although one would normally use an actual data base ID). The second line would contain the name of the sequence (this is optional), and the rest of the file would the sequence itself. An asterisk then ends the entry for that sequence. For example, here is alpha crystallin B-chain in this format:

```
> P1;CRAB_ANAPL

ALPHA CRYSTALLIN B CHAIN (ALPHA(B)-CRYSTALLIN).

MDITIHNPLI RRPLFSWLAP SRIFDQIFGE HLQESELLPA SPSLSPFLMR

SPIFRMPSWL ETGLSEMRLE KDKFSVNLDV KHFSPEELKV KVLGDMVEIH

GKHEERQDEH GFIAREFNRK YRIPADVDPL TITSSLSLDG VLTVSAPRKQ

SDVPERSIPI TREEKPAIAG AQRK*
```

The results can then be viewed after a short wait. As an example, we can find the human, mouse, bovine and pig versions of the protein IL-7 in the UNIprotKB database and prepare a file that contains the sequences to find a multiple alignment. Since there are multiple entries for the IL-7 protein in some species (human and mouse), here we select some arbitrarily. The result is the following file (using the UNIprotKB database identifiers as the database identifiers here),

```
>P1;Q5FBY9

MFHVSFRYIF GLPPLILVLL PVASSDCDIE GKDGKQYESV LMVSIDQLLD

SMKEIGSNCL NNEFNFFKRH ICDANKVKGR KPAALGEAQP TKSLEENKSL

KEQKKLNDLC FLKRLLQEIK TCWNKILMGT KEH*

>P1;Q9N2G6

MFHVSFRYIF GIPPLILVLL PVASSDCDIE GKDGGVYQNV LMVSIDDLDR

MIDFDSNCLN NEPNFLKKHS CDDNKEASFL YRAARKLKQF IKMNISEEFN

HHLSTVSQGT LTLFNCTSKV KGRKPPSLGE AQLTKNLEEN KSLKEQKRQG

DLCFLKILLQ KIKTCWNKIL RGAKEY*

>P1;P10168

MFHVSFRYIF GIPPLILVLL PVTSSECHIK DKEGKAYESV LMISIDELDK

MTGTDSNCPN NEPNFFRKHV CDDTKEAAFL NRAARKLKQF LKMNISEEFN

VHLLTVSQGT QTLVNCTSKE EKNVKEQKKN DACFLKRLLR EIKTCWNKIL

KGSI*

>P1;P26895

MFHVSFRYIF GIPPLILVLL PVASSDCDIS GKDGGAYQNV LMVNIDDLDN

MINFDSNCLN NEPNFFKKHS CDDNKEASFL NRASRKLRQF LKMNISDDFK

LHLSTVSQGT LTLLNCTSKG KGRKPPSLSE AQPTKNLEEN KSSKEQKKQN

DLCFLKILLQ KIKTCWNKIL RGIKEH*
```

whose contents we can simply paste into a web interface. After submitting the job and a short wait, the program returns the following alignment

```
CLUSTAL 2.0.8 multiple sequence alignment

Q5FBY9  NNEFNFFKRHICDAN-------------------------------

P26895  -NEPNFFKKHSCDDNKEASFLNRASRKLRQFLKMNISDDFKLHLSTVSQ

Q9N2G6  -NEPNFLKKHSCDDNKEASFLYRAARKLKQFIKMNISEEFNHHLSTVSQ

P10168  -NEPNFFRKHVCDDTKEAAFLNRAARKLKQFLKMN-------------
```

```
Q5FBY9 ----------KVKGRKPAALGEAQPTKSLE-----ENKSLKEQKKLNDL

P26895 GTLTLLNCTSKGKGRKPPSLSEAQPTKNLE-----ENKSSKEQKKQNDL

Q9N2G6 GTLTLFNCTSKVKGRKPPSLGEAQLTKNLE-----ENKSLKEQKRQGDL

P10168 ----------ISEEFNVHLLTVSQGTQTLVNCTSKEEKNVKEQKK-NDA


Q5FBY9 CFLKRLLQEIKTCWNKILMGTKEH

P26895 CFLKILLQKIKTCWNKILRGIKEH

Q9N2G6 CFLKILLQKIKTCWNKILRGAKEY

P10168 CFLKRLLREIKTCWNKILKGSI--
```

Note that the algorithm finds a local alignment that starts roughly 60 amino acids into the sequences and stretches to the end of the sequences.

11.4 Hidden Markov models

One can use hidden Markov models to conduct an alignment by supposing that there is some unknown model sequence. This sequence is the common sequence that underlies all the observed sequences. The hidden process is the path through the dynamic programming matrix for aligning each observed sequence to the model sequence. This process can be considered as a 2 dimensional Markov process. One dimension of this process is if there is an insertion or a match in the observed sequence (a binary variable) and the other dimension is the number of deletions up to the current location. It is the second component of this 2 dimensional process that necessarily possesses the Markov property, however, for some models the first would be Markov too, as described below. Typically it is assumed that if the hidden process is in a match state, then the distribution for the observed sequence is just a Dirac delta function at the letter of the model state. If the hidden process is in the insertion state then the distribution for the observed sequence is typically modeled as an observation from a single background distribution. This background distribution is parameterized by a set of model parameters that must be estimated. The extreme flexibility of hidden Markov models for multiple alignment comes from the fact that one can specify more complex models for the background distribution (e.g. the background distribution depends on location) and the distribution of the observed sequence given the hidden process is in the match state (e.g. one can allow a general probability distribution for the observations that differs depending on location in the model state). Many researchers have explored the use of HMMs for sequence analysis, including Krogh et al. (1994) and Baldi et al. (1994). The text of Durbin et al. (1998) provides a lengthy discussion and provides many references to the literature. The approach described here is based on Liu, Neuwald and Lawrence (1999).

As a concrete example, consider how an observed sequence could arise given a model

Table 11.1 *Values of the hidden process for generating an observed sequence given a model sequence.*

$\theta_{1,t}$	0	1	1	0	1	1
$\theta_{2,t}$	0	0	0	1	1	1

sequence and values of the hidden process. Although one usually uses protein sequences in the context of this approach, it is also suitable for DNA, which is what we will consider for this simple example. Suppose that the model sequence is just

AACACAA

and we suppose that if the hidden state is in the match state, then the observed sequence has the letter that appears in the model sequence. Hence if $\theta_{1,t} = 0$ then there is an insertion in the observed sequence and if $\theta_{1,t} = 1$ then there is a match while $\theta_{2,t}$ counts the number of deletions up until location t. If the values of $\theta_{1,t}$ and $\theta_{2,t}$ are as given in Table 11.1, then the observation sequence would be

-ACCAA

where the symbol – represents a blank. To fill in this blank value we could suppose that all blanks are filled in iid from a distribution that puts a certain probability on each of the 4 possible letters (this only entails 3 parameters since the 4 probabilities sum to 1). Hence we can specify $p(y_t|\theta_{1,t}, \theta_{2,t})$ via the parameters for the background distribution (3 parameters) and the model sequence (here 7 parameters where each takes 1 of 4 values).

Notice that this data generation mechanism is such that the length of the generated sequence will be a random variable that depends on the sequence of $\theta_t = (\theta_{1,t}, \theta_{2,t})$. That is, prior to knowing θ_t, one can't determine the length of the sequence, hence it will be a random variable. However, once the sequence of θ_t is known, the length of the sequence will be determined since the sequence ends when the last letter in the model sequence is matched to the generated sequence. This random length is referred to as a stopping time in the stochastic process literature, and an important feature of stopping times is that a Markov chain, X_t (defined for all integers $t \geq 1$) that stops at a stopping time, T (a positive integer), is a Markov chain with parameter index $1 \leq t \leq T$. Hence the variable length sequences that are generated in the above fashion are still hidden Markov models in the technical sense.

Now we will consider how to write explicit forms for the conditional distributions that describe the hidden Markov model for a very simple model. Suppose that we code our model sequence via vectors of zeros and ones where only 1 element of the vector is 1. For example, for DNA sequences we could use $(1, 0, 0, 0)$ to represent an A, $(0, 1, 0, 0)$ to represent a C and so on. We will suppose that our sequence data y_t has been coded using this system. Then let ϕ_i represent the ith element of the model sequence (a 4-vector of zeros and ones), a parameter we seek to estimate. Here we suppose that the length of the model sequence, denoted J is known, although there are methods for obtaining estimates of this. We may set J to be small if we seek local alignments and take J to be large if we are seeking a global alignment. In Baldi et al. (1994), the authors set J equal to the average length of the sequences one is aligning so that this would be a global method. Examination of the results from a heuristic technique, such as CLUSTALW, might also help suggest reasonable values for J. Let π_{0i} represent the probability that i^{th} possible sequence value is generated by the background distribution for $i = 1, \ldots, 4$ here. Furthermore, let π_0 represent a vector of these probabilities and let us use the convention that if x and y are vectors with elements x_i and y_i then $x^y = \prod_i x_i^{y_i}$. Then, as the previous example shows, we can express $p(y_t|\theta_t)$ as

$$p(y_t|\theta_t) = \delta^{\theta_{1,t}}_{\{y_t = \phi_{t+\theta_{2,t}}\}} \left(\prod_i \pi_0^{y_t} \right)^{1-\theta_{1,t}},$$

where $\delta_A = 1$ if A is true and is zero otherwise. To specify the transition distribution for the hidden chain $p(\theta_{1,t}, \theta_{2,t}|\theta_{1,t-1}, \theta_{2,t-1})$, we need to specify at least the probability of having a match

$$p(\theta_{1,t} = 1, \theta_{2,t} = i|\theta_{1,t-1}, \theta_{2,t-1} = i),$$

and the probability of a deletion given the current number of deletions

$$p(\theta_{1,t}, \theta_{2,t} = i + 1|\theta_{1,t-1}, \theta_{2,t-1} = i).$$

The simplest model could allow for deletions to occur in an iid fashion.

11.4.1 Extensions

Note that this is a minimal specification of the model in that we have supposed that there is a single probability that governs if there is an insertion or match. By introducing dependence in the transition probabilities for $\theta_{1,t}$ we can make the probability of an insertion depend on if the chain is currently in a match state and thereby have the model allow different penalties for opening versus extending a gap in the model sequence. In addition, we have supposed that all matches count equally in our multiple alignment, an assumption we didn't make when considering pairwise alignments. Finally, we don't need to assume that given that the state process is in a "match state" it will actually produce a match. For example, we could suppose that given the state process is in a "match state," there will be a match between the model sequence and the generated sequence with probability less than one and there will be a mismatch (i.e. the generated sequence will deviate from the model sequence at the current location) with some probability. In the context of aligning amino acid sequences, we could suppose that $y_t|\theta_{1,t}$ is multinomial where the set of multinomial success probabilities depend on the value of $\phi_{t+\theta_{2,t}}$. In this way our model could allow certain mutations to be more likely, which is what we assumed in the context of pairwise alignment. Henikoff and Henikoff (1996) have explored the use of Dirichlet priors to incorporate information from known substitution matrices into the analysis.

The previously described data generating process can be extended in many ways in addition to those just described. For example, we may suppose that there is Markov dependence in the background distribution used to fill in the insertions in our generated sequence. In addition, we may allow the Markov chain to be inhomogeneous in that the transition probabilities depend on location in the sequence. Moreover one can even allow for Markov dependence when describing transitions between the different regions in a inhomogeneous Markov chain (so a hidden Markov model would describe when different hidden Markov models operate).

Given a model for $p(y_t|\theta_t)$ and $p(\theta_t|\theta_{t-1})$ we can use the approach to parameter estimation described in Chapter 9 to obtain the posterior mode of all parameters. An important difference between the use of hidden Markov models described there and the current application is that the model sequence is a parameter that exists on a finite parameter set, the collection of all sequences of length J. For small J one could consider all possible model sequences, but for even moderate J such an approach isn't feasible and one needs to consider other approaches. Another difference here is that since we have multiple sequences, the joint likelihood of all parameters is given by the product over all sequences since we model the sequences as independent given the model parameters. This would allow us to estimate the most likely model sequence and the set of most likely alignments for each sequence (using the Viterbi algorithm for each sequence and the posterior mode for all other parameters). A number programs have been developed for using hidden Markov models to conduct a multiple alignment and are widely available (for example, SAM).

11.5 Block motif methods

Another approach to multiple alignment is based on the idea behind PSSMs. In this approach we suppose that there is some regulatory element described by a PSSM, but we don't know the start location of the PSSM in each sequence. The goal is to determine the start location of the PSSM in each sequence and determine the probabilities of observing the different nucleotides (or amino acids) at the various locations in the PSSM. While many refinements have been introduced since these methods were first proposed, we will first discuss the most simple model and then consider extensions. A substantial literature has been built around these models, but for early references consult Lawrence and Reilly (1990), Cardon and Stromo (1992), Liu (1994) and Liu, Neuwald and Lawrence (1995). These methods are inherently methods of local alignment since they search for small conserved regions in the sequences.

Suppose there are N sequences, $y_1, \ldots, y_N$ over an alphabet of size p and the ith sequence is of length n_i. Denote the collection of these sequences as y. Suppose there is a conserved sequence of length J. A *motif* is the probability distribution for the elements (so a motif is the expected value of a PSSM). Originally, these methods have assumed that biopolymers are sequences of iid draws from some probability distribution, but there have been extensions that depart from this assumption. There are 2 such distributions when we seek a single motif: a background distribution describing the distribution over the letters in non-motif regions and a probability distribution over the motif region. For the motif region, we assume the distribution depends on location in the motif. Hence we can think of a motif as a set of probability distributions over the alphabet $\theta_j = (\theta_{1j}, \ldots, \theta_{pj})$ for $j = 1, \ldots, J$. We use the notation Θ to represent all of these parameters. In addition there is a background distribution $\theta_0 = (\theta_{10}, \ldots, \theta_{p0})$. In this context, an alignment is simply the location of the start of the motif in each sequence $\alpha = \{(i, \alpha_i) : i = 1, \ldots, N\}$, where α_i is the location of the start of the motif in the ith sequence.

If we treat all of the parameters and the alignment, α, as random variables, then we can use methods of Bayesian statistics to find the posterior distribution of the alignment. Now $p(\alpha|y) \propto p(y|\alpha)p(\alpha)$ thus if we suppose that $p(\alpha) \propto 1$ (i.e. all alignments are equally likely in the prior)

$$p(\alpha|y) \propto \int p(y|\alpha, \theta_0, \Theta)p(\theta_0, \Theta)\, d\theta_0 d\Theta. \qquad (11.1)$$

If we use independent, conjugate prior distributions for the parameters θ_0 and Θ then we can perform the desired integration. Our likelihood is a product of multinomial distributions, hence the conjugate priors for each of the parameter vectors θ_j for $j = 0, \ldots, J$ is the Dirichlet distribution. Conditional on a vector of parameters β_j, if we introduce the notation that for 2 vectors x and y, $\Gamma(x) = \prod_i \Gamma(x_i)$, $|x| = \sum_i x_i$, and $x^y = \prod_i x_i^{y_i}$, then the density function of this distribution can be expressed

$$p(\theta_j|\beta_j) = \frac{\Gamma(|\beta_j|)}{\Gamma(\beta_j)}\theta_j^{\beta_j - 1}.$$

To write the posterior in a simple fashion, we introduce some notation. Define the function h to map a set of letters to a p vector that counts the frequency of each letter. We also introduce a bracket notation that takes a single starting position of an element and returns the entire element, i.e. $\{\alpha\} = \{(k, \alpha_i + j - 1) : i = 1, \ldots, N, j = 1, \ldots, J\}$. In addition we define the set $\alpha(j) = \{(k, \alpha_i + j - 1) : i = 1, \ldots, N\}$. With these definitions in addition to our rules for powers of vectors given above, we can write

$$p(y|\alpha, \theta_0, \Theta) \propto \theta_0^{h(y_{\{\alpha\}^c})} \prod_{j=1}^{J} \theta_j^{h(y_{\alpha(j)})}.$$

Using this in equation 11.1 in addition to our priors, we find that

$$p(\alpha|y) \propto \int \theta_0^{h(y_{\{\alpha\}^c})} \prod_{j=1}^{J} \theta_j^{h(y_{\alpha(j)})} \prod_{j=0}^{J} \theta_j^{\beta_j - 1}\, d\theta_0 d\Theta$$

$$\propto \int \theta_0^{h(y_{\{\alpha\}^c}) + \beta_0 - 1} \prod_{j=1}^{J} \theta_j^{h(y_{\alpha(j)}) + \beta_j - 1}\, d\theta_0 d\Theta$$

$$\propto \frac{\Gamma(h(y_{\{\alpha\}^c}) + \beta_0)}{\Gamma(|h(y_{\{\alpha\}^c}) + \beta_0|)} \prod_{j=1}^{J} \frac{\Gamma(h(y_{\alpha(j)}) + \beta_j)}{\Gamma(|h(y_{\alpha(j)}) + \beta_j|)}$$

hence

$$p(\alpha|y) \propto \Gamma(h(y_{\{\alpha\}^c}) + \beta_0) \prod_{j=1}^{J} \Gamma(h(y_{\alpha(j)}) + \beta_j),$$

since

$$\Gamma\left(|h(y_{\{\alpha\}^c}) + \beta_0|\right) = \Gamma\left(\sum_{i=1}^{N} n_i - JN + |\beta_0|\right) \tag{11.2}$$

and

$$\Gamma(|h(y_{\alpha(j)}) + \beta_j|) = \Gamma(N + |\beta_j|)$$

neither of which depend on α.

While we have an expression for the posterior distribution of interest, it is not clear what we should do with this expression in order to conduct inference. Note that our parameters of primary interest are just the start locations of the motif (namely α_i for $i = 1, \ldots, N$) and as such are integer valued. The approach of Liu et al. (1995) is to find the conditional distribution of α_i given $\alpha_{[-i]} = \{(j, a_j) : j \neq i\}$, then iteratively sample from these conditional distributions, or approximations to these distributions. That is, they propose to use the Gibbs sampler in order to conduct the inference. As an alternative, one could use the EM algorithm to iteratively locate the start sites (which plays the role of the missing data), then find the most likely values of the probabilities describing the motif. Here we will show how compute the conditional distributions of the start sites given the data (the E-step of the EM algorithm would use these conditional distributions to compute the expected value of the start site). The EM based approach is implemented in the software program called MEME and been extended in many ways similar to the model extensions we discuss below.

Note that the conditional distribution of α_i is given by the following

$$p(a_i|\alpha_{[-i]}, y) = \frac{p(\alpha|y)}{p(\alpha_{[-i]}|y)},$$

but

$$p(\alpha_{[-i]}|y) \propto \Gamma(h(y_{\{\alpha_{[-i]}\}^c}) + \beta_0) \prod_{j=1}^{J} \Gamma(h(y_{\alpha_{[-i]}(j)}) + \beta_j)$$

since we just ignore the motif element in sequence i to get this conditional. Next, note that the bracket notation implies $\{\alpha_i\} = \{(i, \alpha_i + j - 1) : j = 1, \ldots, J\}$, hence

$$h(y_{\{\alpha_i\}}) + h(y_{\{\alpha\}^c}) = h(y_{\{\alpha_{[-i]}\}^c}), \tag{11.3}$$

so that

$$p(\alpha_i | \alpha_{[-i]}, y) \quad \propto \quad \frac{\Gamma(h(y_{\{\alpha\}^c}) + \beta_0)}{\Gamma(h(y_{\{\alpha_{[-i]}\}^c}) + \beta_0)} \prod_{j=1}^{J} \frac{\Gamma(h(y_{\alpha(j)}) + \beta_j)}{\Gamma(h(y_{\alpha_{[-i]}(j)}) + \beta_j)}$$

$$\propto \quad \frac{\Gamma(h(y_{\{\alpha_{[-i]}\}^c}) - h(y_{\{\alpha_i\}}) + \beta_0)}{\Gamma(h(y_{\{\alpha_{[-i]}\}^c}) + \beta_0)} \prod_{j=1}^{J} \frac{\Gamma(h(y_{\alpha(j)}) + \beta_j)}{\Gamma(h(y_{\alpha_{[-i]}(j)}) + \beta_j)}$$

$$\propto \quad \frac{\Gamma(h(y_{\{\alpha_{[-i]}\}^c}) - h(y_{\{\alpha_i\}}) + \beta_0)}{\Gamma(h(y_{\{\alpha_{[-i]}\}^c}) + \beta_0)}$$

$$\times \quad \prod_{j=1}^{J} \frac{\Gamma(h(y_{\alpha_{[-i]}(j)}) + h(y_{i,\alpha_i+j-1}) + \beta_j)}{\Gamma(h(y_{\alpha_{[-i]}(j)}) + \beta_j)}$$

where the last line is due to the equality

$$h(y_{\alpha(j)}) = h(j_{\alpha_{[-i]}(j)}) + h(y_{i,\alpha_i+j-1}).$$

In this last expression, $h(y_{i,\alpha_i+j-1})$ is just a p vector with a single 1 and the rest of the entries zero. But if m is a vector of positive integers and v is a vector of all zeros except a 1 at a single entry, then

$$\Gamma(m + v)/\Gamma(m) \quad = \quad \prod_{i} \Gamma(m_i + v_i)/\Gamma(m_i)$$

$$= \quad \prod_{i} (\Gamma(m_i + v_i)/\Gamma(m_i))^{v_i}$$

$$= \quad \prod_{i} m_i^{v_i}$$

$$= \quad m^v$$

where the second to last line is by the relationship between the factorial function and the gamma function. Hence we arrive at the conditional

$$p(\alpha_i | \alpha_{[-i]}, y) \quad \propto \quad \frac{\Gamma(h(y_{\{\alpha_{[-i]}\}^c}) - h(y_{\{\alpha_i\}}) + \beta_0)}{\Gamma(h(y_{\{\alpha_{[-i]}\}^c}) + \beta_0)}$$

$$\times \quad \prod_{j=1}^{J} [h(y_{\alpha_{[-i]}(j)}) + \beta_j]^{h(y_{i,\alpha_i+j-1})}.$$

Since this expression involves the ratio of 2 gamma functions, the authors actually use an approximation in their computations. This approximation uses the trick about ratios of gamma functions given above, which is exact if the vector v has just a single non-zero entry and is an approximation in other circumstances. Thus they approximate the ratio of gamma functions with $[h(y_{\{\alpha_{[-i]}\}^c}) + \beta_0]^{-h(y_{\{\alpha_i\}})}$, and this gives the approximate posterior as

$$p(\alpha_i | \alpha_{[-i]}, y) \propto \prod_{j=1}^{J} \left[\frac{h(y_{\alpha_{[-i]}(j)}) + \beta_j}{h(y_{\{\alpha_{[-i]}\}^c}) + \beta_0} \right]^{h(y_{i,\alpha_i+j-1})},$$

or more simply

$$p(\alpha_i = k | \alpha_{[-i]}, y) \quad \propto \quad \prod_{j=1}^{J} \left(\frac{\hat{\theta}_{j[-i]}}{\hat{\theta}_{0[-i]}} \right)^{h(y_{i,k+j-1})}, \tag{11.4}$$

for $k = 1, \ldots, n_i - J + 1$ where $\hat{\theta}_{j[-i]}$ is the vector of posterior means for position j given the current alignment, and $\hat{\theta}_{0[-i]}$ is the vector of posterior means for the locations not involved in the alignment. Since we have an expression for something proportional to $p(\alpha_i = k | \alpha_{[-i]}, y)$ for each position, k, in sequence y_i we can draw samples from this distribution since there are only a finite number of possible values for α_i. Hence the algorithm proceeds by sampling from each of these distributions iteratively. A simple way to sample from a finite set (as here) is to simulate u from a uniform distribution on (0,1), then set $\alpha_i = k$ so that

$$\sum_{\ell=1}^{k-1} p(\alpha_i = \ell | \alpha_{[-i]}, y) < u \leq \sum_{\ell=1}^{k} p(\alpha_i = \ell | \alpha_{[-i]}, y).$$

Recall that as one samples the vector of start sites, $(\alpha_1, \ldots, \alpha_N)$, using the Gibbs sampler, as soon as a start site is sampled on the ℓth iteration, say α_i^ℓ, one then uses this sampled value to sample α_j on the ℓth iteration for $j > i$.

11.5.1 Extensions

There are several important extensions that one can build into the basic model. These extensions are very important from the perspective of applications, but are straight-forward extensions from the perspective of finding the conditional distributions. First, we can easily accommodate multiple instigations of the same motif within a given sequence. The trick is to pretend all of the sequence data is just from one long sequence and look for motifs that don't overlap or extend over boundaries of the actual sequences. In practice one can just ignore draws from the Gibbs sampler that have these properties.

Slightly more complicated is the extension to insertions in the motifs. Here we select a number $W > J$ and look for a motif of length W. The difference is, we now introduce a set of W indicator variables that indicate if each position is part of the motif (so only J of these indicators are 1 and the rest are zero). One must provide a prior for this vector. In practice a uniform prior over the space of $\binom{W}{J}$ indicator variables that have J ones seems too diffuse, hence priors that restrict the span of the motif are typically used. These models are referred to as *fragmentation models*.

Similarly, we can treat the case where not all sequences have a motif element and some sequences have multiple motif elements. The idea is to once again treat all the sequences as one long sequence of length L^* and introduce $L^* - J + 1$ indicator variables that represent if an element starts at each possible location. In addition, we can even allow for these indicator variables to be n valued instead of just binary. In this way we can treat the problem of more than one type of motif.

11.5.2 The propagation model

The primary drawback of the block motif methods is that one must specify the size of the blocks and the number of motifs. In contrast, the hidden Markov model based approaches (discussed in Section 11.4) do not generally require the user to specify this information (although some implementations do specify this), but the generality of hidden Markov models for multiple alignment implies they are often not very sensitive due to the large number of parameters that must be estimated (Grundy et al., 1997). Hence there is a desire for methods that can are sensitive yet flexible, requiring little user input. The propagation model of Liu, Neuwald and Lawrence (1999) seeks to fulfill these requirements. This method has been implemented in the software called PROBE.

Consider the basic set up of the block motif methods. Suppose there are L elements in each sequence, and use α to now denote the start locations of all of these elements in each sequence. Let w_l denote the width of the lth motif and suppose the probabilities governing the frequencies of the letters in this motif are given by θ^ℓ. If $\alpha_{.l}$ is the starting position of the lth motif in all sequences, then supposing there are no deletions in the motifs, we can write the posterior for α under a flat prior (or the likelihood for y) as

$$p(\alpha|y,\theta_0,\Theta) \propto \theta_0^{h(y)} \prod_{l=1}^{L} \prod_{j=1}^{w_l} (\theta_j^l/\theta_0)^{h(y_{\alpha_{.l}+j-1})}.$$

As before, we can integrate θ_0 and Θ out of the posterior to obtain the marginal posterior for the alignment given the sequence data as

$$p(\alpha|y) \propto \Gamma(h(y_{\{\alpha\}^c}) + \beta_0) \prod_{l=1}^{L} \prod_{j=1}^{w_l} \Gamma(h(y_{\alpha_{.l}+j-1}) + \beta_j^l).$$

The novel aspect of these models that allows for a connection to the hidden Markov models is that we now use a Markovian prior for the distribution of the motif start sites within each sequence:

$$p(\alpha_{i.}) \propto \prod_{l=1}^{L-1} q_l(\alpha_{i,l}, \alpha_{i,l+1}),$$

where $\alpha_{i.}$ is the portion of the alignment relevant to sequence i. We suppose the alignments are independent across the different sequences $p(\alpha) = \prod_i p(\alpha_{i.})$. If $w_l = 1$ for all l then this model reduces to the usual hidden Markov model with no deletions from the model sequence and insertions with lengths indicated by the differences between the $\alpha_{i,l}$. For the function $q_l(x,y)$ we require $q_l(x,y) = 0$ for $|x-y| < w_l$ for motifs to not overlap. While any form of q that satisfies this criterion is acceptable, Liu, Neuwald and Lawrence (1999) recommend using $q = 1$ when it is not zero. Thus our prior for the start locations of the motifs in sequence i can be described as being uniform over the set of all start locations that don't violate the rule that a motif can't start until the previous motif has ended.

The computation of the posterior distribution mimics the calculations for the block motif model. Once again, we compute the conditional distributions and use the Gibbs sampler to simulate from the posterior distribution. We find

$$p(\alpha_{i\cdot} = (k_1, \ldots, k_L) | y, \alpha_{[-i]}) \quad \propto \quad \left[\prod_{l=1}^{L-1} q_l(k_l, k_{l+1}) \prod_{j=1}^{w_l} \left(\frac{\hat{\theta}_j^l}{\hat{\theta}_0} \right)^{h(y_{i,k_l+j-1})} \right]$$

$$\times \quad \prod_{j=1}^{w_L} \left(\frac{\hat{\theta}_j^L}{\hat{\theta}_0} \right)^{h(y_{i,k_L+j-1})},$$

where $\hat{\theta}$ are posterior means as before. A complication that now arises is that we need the normalizing constant for this conditional posterior in order to sample. This is because we only specified our prior so that all possible start locations of the motifs are equally likely. If we can determine the total number of possible start locations for all motifs in one sequence that are compatible with the requirement that they don't overlap, say there are M, then we would be able to specify a prior for all possible start sites that supposes they all have probability $\frac{1}{M}$. The authors provide a recursive algorithm in order to compute this normalizing constant thus allowing us to use the Gibbs sampler to sample start locations as before.

As in the usual block motif model, we can consider the extension of the fragmentation model and allow for deletions as before. Also as in the block motif model, one must specify the block sizes, w_l, and number of motifs, L. The authors propose to use *Bayes factors*, which involve ratios of $p(y|W, L)$ for various choices of W and L to select these values, but since one cannot compute this probability in any simple way (one could use MCMC as described in Carlin and Louis, 2009) they propose to use an approximate method based on $p(y|\hat{\alpha})p(\hat{\alpha})$, where $\hat{\alpha}$ is the posterior mode of the alignment. In practice often several values are examined for these parameters and the results are inspected. These methods continue to be extended in a variety of ways (see, for example, Thijs et al., 2001; Webb, Liu and Lawrence, 2002; Thompson, Rouchka and Lawrence, 2003) and many programs are now available, such as BIO-PROSPECTOR and ALIGNACE, that allow implementation of these approaches.

11.6 Enumeration based methods

The statement of the problem of finding a multiple alignment using enumeration methods differs from the stochastic approaches described above. Instead of a motif as a probability distribution, we seek to find a sequence of length J that is found in all N input sequences where J is as large as possible, and differences at positions are allowed, but penalized. This approach is similar to the BLAST and FASTA algorithms in that exact matches are found between the sequences by hashing the sequences. By finding all k-tuples and saving the information in a hash table, we can easily detect subsequences of length k that are common across all of the sequences. Moreover, if we keep track of how many mismatches are found as we examine if the k-tuple is present we can allow mismatches provided they stay below a user specified bound.

In addition, we can drop subsequences from consideration sequentially. For example, we could first see if an A is present in all of the sequences, then AA, then AAA, and if AAA is not present we needn't consider any k-tuples for $k \geq 3$ that contain an AAA. This was the approach first described by Sagot (1998), and while useful, is still too slow to be of practical use for finding motifs of reasonable size in DNA sequences of typical length. Nonetheless, a number of approaches have been developed that examine hash tables for small k-mers (such as $k = 12$) and look for k-mers that are similar and found among the hash tables of most sequences. Frequently such methods utilize some clustering method to find groups of nearly similar k-mers that are found in multiple sequences (an example of such a method is described in section 11.7).

Due to the prohibitive computational requirements of enumerating every subsequence that is common to a set of sequences (or sufficiently close), a number of refinements have been made to this method. Another method due to Pavesi, Giancarlo and Pesole (2001) that avoids the prohibitive computational burden of exhaustive searches is consideration of only sequences that are sufficiently close as one searches for common subsequences. For concreteness, suppose that we will allow 25% of the positions to differ across the sequences. We then determine if a sequence is sufficiently close by noting that when we consider a sequence of length 16 say, this will be composed of 4 subsequences of length 4, and each of these subsequences will have errors. Thus we don't consider 16-tuples that are composed of 4-tuples where each of the 4-tuple has more than 1 error. Note that such an approach would not consider 16-tuples with 3 errors in the first 8 positions. In this sense, this algorithm searches for sequences where the errors are spread over the entire length of the sequence. Note also that, as described here, the presence of insertions and deletions could have a serious impact since a single insertion near the start of the motif would lead to many pointwise errors. Various other modifications to the basic exhaustive search approach and freely available software (called WEEDER) is available to implement the algorithm.

11.7 A case study: detection of conserved elements in mRNA

While most focus on the regulation of gene expression through the transcriptional stage of gene expression, many researchers are starting to focus on post-transcriptional mechanisms that alter gene expression (see, for example, Barreau et al., 2006; Moraes, Wilusz and Wilusz, 2006). Recall that once an mRNA molecule has been produced, it enters the cytosol and interacts with proteins, protein complexes and ribosomes. An mRNA molecule will eventually decay (i.e. break apart) in the cytosol, however the mRNA products of different genes can have vastly different lifespans in complex organisms. If an mRNA molecule decays rapidly, it is unlikely to produce much of its corresponding protein product. Hence via regulation of the decay of an mRNA molecule, the cell is able to alter the gene expression for a gene. One widely studied method of post-transcriptional regulation via mRNA decay is the regulation of decay by AU-rich elements (AREs). AREs are sequences found in the 3' UTR of many genes that have short lived mRNA transcripts. While a number of subclasses of AREs have been described, the basic feature of the sequence is the subsequence

AUUUA (Chen et al., 2001). AREs are thought to operate by interacting with several proteins that help recruit the mRNA decay machinery to the location of the mRNA molecule. While much is known about AREs, they fail to account for the rapid decay of many mRNA molecules, hence similar control sequences are thought to exist.

In one attempt to find these other regulatory sequences, Vlasova et al. (2008) measured gene expression over time for many genes and used motif finding algorithms to identify new regulatory sequences in the 3' UTR of rapidly decaying genes. These experiments used human T cells to estimate the decay rates for many genes simultaneously. These cells are essential for normal functioning of the immune system and are quite dynamic, active cells relative to many cells in fully developed humans. To identify rapidly decaying genes, the authors first exposed the collection of cells to a compound that stops gene transcription (Actinomycin D), then measured the level of gene expression at 4 time points (the experiment was replicated 3 times). Since transcription is stopped at the baseline time point, any mRNA molecules that are detected in the sample at time t must have been produced prior to the baseline time point. Moreover, by using the data from all 4 time points and replicates, one can use linear regression to estimate the rate of decay in gene expression for each gene separately. If y_{ijk} represents the level of expression for gene i, time point j and replicate k, then a simple model for the decay process is to assume that

$$\mathrm{E}y_{ijk} = \beta_{i0}\mathrm{e}^{-\beta_{i1}t_j}$$

for 2 gene level parameters β_{i0} and β_{i1}. Such a model is referred to as a first order decay model. If we take the logarithm in this expression, we have a model that supposes the mean of the level of gene expression is a linear function of time, hence we can use linear regression to obtain estimates of the parameters in the model. Just using linear regression on the measured mRNA levels, one can find hundreds of transcripts that have rapid decay (Raghavan et al., 2002). While the original publication used a complex nonlinear model to describe the decay process due to the use of certain tools for measuring gene expression, the results described here used a simple linear regression estimate to obtain decay rates from more up to date methods for summarizing gene expression measurements from a microarray. We then used these linear regression estimates of the decay rate to compute the halflife of each mRNA transcript. The halflife of a molecule is the time it takes until half of the transcript has decayed. It is not hard to show that if $\tau_{\frac{1}{2}}$ is the halflife of a gene whose decay follows the first order decay model, then $\tau_{\frac{1}{2}} = \log(2)/\beta_1$, so that we can obtain an estimate of the halflife from our linear regression estimates. We classify genes as having short halflifes if the halflife for the gene is less than 60 minutes, whereas a gene with a halflife that exceeds 360 minutes is classified as having a long halflife. Only a small fraction (about 5%) of the short lived transcripts contained an ARE.

We then looked for motifs that were common among the 3' UTR of those genes that we classified as being short-lived. An enumeration based method and a block motif model suggested the presence of very similar motifs that were present in these transcripts. The mode of the motif, which has been named a GU-rich element (GRE), is the following: UGUUUGUUUGU. The enumeration based approach first examined the frequency of all possible 12-mers in the group of short lived transcripts and in the set of long lived transcripts. Following this, a list of the top 1,000 12-mers that had a higher frequency in the short-lived transcripts than the long-lived transcripts was constructed. If we can find 12-mers that are close in this set, we interpret that as resulting from different instigations of a common motif. So, for each 12-mer in our set of 1,000, we compute how similar it is to every other 12-mer in the set. To determine the level of similarity, we just use the proportion of bases that match for each pair. Then hierarchical clustering is used to find groups of 12-mers that exhibit a large amount of agreement. Note that this approach uses information from the long-lived transcripts to help screen out 12-mers that occur more commonly in 3' UTR sequences regardless of the halflife of the transcript. A Gibbs sampling based approach to block motif methods, in addition to an implementation for that model based on the E-M algorithm, also found evidence that this is a motif that is present in the short-lived transcripts.

Once the mode of the motif was found, further instigations were found in the 3' UTR of other genes. To this end, 4812 transcripts were scanned for the presence of the GRE. If we seek exact matches or those that allow one mismatch, we find that this motif is present in roughly 5% of all genes for which we have expression data. This is comparable to the situation with AREs. Moreover there is no tendency for transcripts to have AREs and GREs so that this appears to be an independent mechanism for regulating gene expression. Further experiments were then conducted to determine what proteins bind to this sequence element and assess the functional impact of the sequence. These experiments showed that it is CUG-binding protein 1 (CUGBP1) that binds to this sequence element, and by inserting this sequence into the 3' UTR of a stable transcript, one can make it decay rapidly. Thus here computational approaches were able to play a key role in the overall experimental goal of furthering our understanding of the mechanisms of gene expression.

11.8 Exercises

1. Use CLUSTALW to find a multiple alignment for the set of defensin proteins found in humans (there are at least 4 listed in the UniProtKB database with accession numbers P59665, P59666, P12838 and Q01524). Now find a multiple alignment using 1 human defensin (use P59665) and defensins from at least 3 other species (for example, P11477 is a mouse defensin and P91793 is a defensin found in the mosquito that transmits yellowfever). Compare the alignments and describe your findings.

2. Show that the result given in equation 11.2, that is,

$$\Gamma\left(|h(R_{\{A\}^c}) + \beta_0|\right) \;\; = \;\; \Gamma\left(\sum_k^K n_k + JK + |\beta_0|\right)$$

is true.

3. Show that the result given in equation 11.3, that is,

$$h(R_{\{a_k\}}) + h(R_{\{A\}^c}) \;=\; h(R_{\{A_{[-k]}\}^c}),$$

is true.

4. Show that the result given in equation 11.4, that is,

$$p(a_k = i | A_{[-k]}, R) \;\propto\; \prod_j \left(\frac{\hat{\theta}_{j[-k]}}{\hat{\theta}_{0[-k]}} \right)^{h(r_{k,i+j-1})},$$

is true.

5. Explain why the method presented for sampling from the conditional distribution $p(\alpha_i | \alpha_{[-i]}, y)$ in the context of using the Gibbs sampler for block motif methods is valid (*Hint:* Show that if the random variable X has cdf $F(x)$ then the random variable $Y = F(X)$ is distributed like a uniform random variable on $(0,1)$).

6. The website http://bio.cs.washington.edu/assessment/ has a set of 52 sequences that have been used for testing motif discovery algorithms (see Tompa et al., 2005 for more information on how these sequences were generated). Extract the yeast datasets and use BIOPROSPECTOR and WEEDER to identify motifs. Summarize your findings.

7. Use the TRANSFAC database to identify as many human binding sites and their associated transcription factors (use only those for which TRANSFAC also provides a binding site consensus sequence). In addition, ignore sites with more than one instance, with any missing information and those whose distance from the transcription start site is more than 3000 bp upstream or an distance beyond the transcription start site (so ignore sites that are annotated with reference to anything other than the transcription start site).

   (a) Generate background sequences using a third order Markov model fit to intergenic human DNA sequences, and randomly insert the binding sites into the generated sequences. Then use MEME, ALIGNACE and WEEDER to identify the sites. Describe in detail the specifics of your implementation (e.g. what is average length of the sequences use) and summarize your findings.

   (b) Repeat the previous part, but instead of using randomly generated background sequences, use the real sequence for the binding site.

CHAPTER 12

Statistical Genomics

12.1 Functional genomics

Recently there has been great interest in understanding how the entire repertoire of genes in a genome are involved in complex cellular phenomena (for example, Lipshutz et al., 1998; Cheung et al., 1998; Wen et al., 1998; Duggan et al., 1999; Holter et al., 2000; Young, 2000; Zhao, Prentice and Breeden, 2001). The term genomics has been used to describe the study of the genome and functional genomics is an area concerned with how the genome impacts phenotypes of interest. While for many years researchers focused on the action of one or a small set of genes involved in some phenomena, there is increasing interest in how many genes work together to produce phenotypes of interest. There have been technological developments that allow researchers to pursue these more sophisticated issues, but these technologies have introduced a host of challenges for data analysts. This has led to the development of many statistical approaches for the array of problems encountered. As in the biopolymer feature identification and discovery literature, the proposed methods have largely been judged on pragmatic considerations such as performance of the algorithm on a widely used test data sets. This is in contrast to the usual statistical evaluation of data analysis methods, and there has been controversy over the role of theoretical statistics for the evaluation of these procedures. This controversy has been fueled by the small sample sizes frequently encountered in these data sets (thereby making asymptotic arguments less salient) and the desire to utilize nonparametric procedures.

We will proceed as follows. After covering the basics of the technology and some basic manipulations of the data that fall beyond image analysis, we will consider the 2 sample problem as it has been formulated in the context of the analysis of microarrays (Chapter 13). A number of specialized approaches have been developed that have been shown to outperform the usual approaches to the 2 sample problem, and we will highlight the commonalities of these methods. Next we will discuss the problem of cluster analysis from the perspective of microarrays (Chapter 14), and we will finally discuss the problem of classification (Chapter 15). Our goal in these discussions will be to proceed in a rigorous fashion in line with usual approaches to the statistical analysis of these problems, and we will consider approaches specialized to the microarray context from this perspective.

12.2 The technology

The past decade has seen the development of powerful new techniques that allow one to measure the level of gene expression for many genes (or more properly, *transcripts*, i.e. mRNA molecules) in a tissue sample simultaneously. These methods use a small chip that has nucleotide sequences bound to the surface of the chip. To understand how this technology works, we will specialize the discussion to the 2 major varieties of microarrays used in practice in the following sections: *spotted cDNA arrays* and *oligonucleotide arrays* (which are also called *tiling arrays*).

The unifying theme of microarray technology is hybridization of labeled samples to a solid surface that has had probes composed of nucleotide sequences bound to it. Suppose we want to determine the quantity of the mRNA corresponding to the gene for mitogen activated protein kinase 9 (a protein found in some human cells) in some set of cells. One would first obtain a subset of the sequence of this gene, such as
GGAATTATGTCGAAAACAGA

and attach many copies (on the order of $1.0e^6$-$1.0e^9$) of it onto a solid surface. Some technologies robotically place these sequences on a glass surface (i.e. spotted cDNA arrays) whereas others simultaneously construct all subsequences for all genes on the array (i.e. oligonucleotide arrays). Then one isolates the RNA from the set of cells. Note that the mRNA corresponding to this DNA sequence would have the following sequence
CCUUAAUACAGCUUUUGUCU.

The next step is to convert the RNA isolated from the cell into complementary DNA (cDNA) so that we obtain the following sequence
CCTTAATACAGCTTTTGTCT.

Note that this sequence is the complement of the sequence that was attached to the chip hence this cDNA will tend to form hydrogen bonds with the sequence we have immobilized on the solid surface and thereby bind to the surface, that is, hybridize to the array. For spotted cDNA arrays (one of the major classes of microarrays), a fluorophore is incorporated into the cDNA molecule so that the relative quantity of the fluorophore can be determined after the hybridization takes place. If one uses multiple fluorophores (usually 2) to label 2 different samples then by comparing how much of each fluorophore is measured after both are mixed and the mixture is hybridized to the array, one can determine the relative quantity of the gene's mRNA in the 2 samples. For the other major variety of microarrays (oligonucleotide or tiling arrays) this cDNA molecule is used to produce cRNA which is then the complement of the original sequence. Typically researchers are interested in measuring the relative amounts of many mRNA molecules simultaneously, so sequences from many genes are bound to the microarray surface. This allows one to measure the relative quantity of many mRNA molecules simultaneously. We will use the term spot to represent a transcript that is measured with a microarray since frequently more than one spot refers to the same transcript. Since there are differences in the construction, use and analysis of the 2 major varieties of arrays, we will now specialize the discussion to accommodate these varieties. Further background information, including detailed experimental protocols can be found in Bowtell and Sambrook (2003).

12.3 Spotted cDNA arrays

Spotted cDNA arrays are most easily distinguished from the other type by the use of competitive hybridization. In this type of array, one first labels each of the 2 samples of interest using 2 distinct fluorophores, or dyes (the 2 types are referred to as Cy3 and Cy5 and represented with the colors green and red respectively). The 2 dyes are also refered to as the 2 channels. Then one mixes the 2 labeled samples together and hybridizes the mixture to the array which has nucleotide sequences for thousands of genes bound to it. This allows one to determine, for each transcript on the array, the relative quantity of that transcript in the 2 samples.

The most basic data for these sorts of arrays are the 2 quantities of mRNA measured for each transcript whose sequence has been placed on the array. Frequently, the first data reduction is to take the ratio of the level of mRNA expression for the 2 samples and use this for analysis. This ratio and its expected value is called the *fold change* across conditions, with the context making clear to which one is referring. Originally, if this ratio exceeds 2 or was less than 0.5 the gene was considered to be differentially expressed, but currently statistical criteria are employed and the problem of determining differential gene expression has been cast as a hypothesis testing problem (Wittes and Friedman, 1999; Hilsenbeck et al., 1999; Ting Lee et al., 2000).

12.4 Oligonucleotide arrays

Oligonucleotide arrays use one or more oligonucleotides, which are relatively short sequences of nucleotides, to measure the quantity of an mRNA molecule in a sample. The most commonly used type (although by no means the only sort in common use) is produced by a company named Affymetrix, and these arrays have been the subject of many investigations using microarrays. Another manufacturer, Agilent, produces oligonucleotide arrays that use a 2 color platform (as in the cDNA arrays described above) but uses oligonucleotide probes; thus the frequent distinction between spotted cDNA arrays and oligonucleotide arrays in terms of competitive hybridization is not quite accurate. Nonetheless, given the widespread use of the arrays produced by Affymetrix and the intense research effort that has been undertaken to optimize the analysis of these arrays, our further discussion of oligonucleotide arrays will focus on these arrays.

Perhaps the greatest distinction between these arrays and the arrays used in the context of the spotted cDNA technology (and other oligo arrays) is the use of multiple probes per transcript and the use of "matched" control probes. For these sorts of arrays, typically 10–20 25-mer probes are used for each transcript (i.e. the probes are 25 nucleotides long). In addition to these so called "perfect match" probes, there are also control probes that are identical to the perfect match probes in terms of nucleotide sequence, except the single nucleotide in the center of the 25-mer is switched to its complementary base (these are called mismatch probes). These distinguishing features have led to a great deal of work in attempting to determine how to best use these data.

12.4.1 The MAS 5.0 algorithm for signal value computation

The algorithm used by the software produced by Affymetrix (the current version is called Microarray Suite 5.0, or simply MAS 5.0) first constructs a quantity called the ideal mismatch based on the mismatch data for each probe, then measures the level of gene expression by using robust averages of the differences in the perfect match and ideal mismatch probes for all of the probes in a the probe set for a transcript. While the construction of robust estimates is desirable, especially in a context where one is summarizing huge amounts of data and can't graphically look for outliers very easily, by throwing out or discounting data from probe pairs with large differences between the perfect match and mismatch probes (which is what a robust method may do), one may underestimate differences in gene expression. In addition to quantifying the level of gene expression for all transcripts on the array, this software also tests the hypothesis that each transcript is expressed in the tissue sample (this is called the presence call). We will now examine these methods in greater detail.

The signal value calculation first estimates a value for downweighting mismatch probe intensities so that the ideal mismatch is less than the perfect match, then computes a robust average of log ratios of the perfect match to the mismatch probe intensities. If we let (PM_i, MM_i) for $i = 1, \ldots, m$ represent the perfect match and mismatch probe intensities for a probe set with m pairs of probes, then we set the ideal mismatch, denoted here IM_i, equal to MM_i if $MM_i < PM_i$ and we set $IM_i = PM_i/\lambda$ for some λ if this condition fails to hold. Note that if $\lambda > 1$ then $PM_i > IM_i$ for all i and so our control probe always has a lower intensity then our test probe. The value of λ is determined by how much larger the perfect probes are than the mismatch probes for this probe set. In greater detail, first set

$$x_i = \frac{PM_i}{MM_i},$$

then compute the mean of these x_i values, denoted $\bar{x}$. Then set

$$v_i = \frac{x_i - \bar{x}}{cS + \epsilon}$$

where S is median of $|x_i - \bar{x}|$, and by default $c = 5$ and $\epsilon = 1.0\mathrm{e}^{-4}$. We then find the weights for each of the x_i, where the weight is zero if $|v_i| \geq 1$ and $w_i = (1 - v_i^2)^2$ else. Finally λ is computed by finding a weighted average of the x_i using these weights, i.e.

$$\lambda = \frac{\sum_i w_i x_i}{\sum_i w_i}.$$

Note that λ is a weighted average of the x_i, so we could run the algorithm again but use the newly computed value of λ in place of $\bar{x}$. In fact this process can be iterated until the value of λ stabilizes at some value-the resulting algorithm produces an estimate known as Tukey's biweight estimate and is a traditional way of estimating the mean of a set of random variables that is robust to measurement errors. Hence the λ value in the MAS 5.0 algorithm is determined by 1 iteration of Tukey's biweight estimate. Note that is is possible for $\lambda < 1$ (see exercise 2). With the ideal mismatches, we can then compute $y_i = \log_2(PM_i - IM_i)$. The signal value, denoted s_i for spot i, is then found by using 1 iteration of Tukey's biweight algorithm using the y_i in place of the x_i as descried above. Finally, the software scales all arrays to a common value by computing a scale factor, denoted f, for each array. This value is found by dividing a target value (such as 500) by the trimmed mean of the exponentiated signal values (exponentiating by 2, i.e. 2^{s_i}). By default, the lowest and highest 2 percent of the signal values are discarded when computing this trimmed mean. The scaled signal value is then given by the $s_i + \log_2 f$.

In addition to the signal value, for each transcript the MAS 5.0 software conducts a test of the hypothesis that the transcript is not expressed in the tissue sample. The basic idea is that if a transcript is not expressed, then the value of the intensities for the perfect match and the mismatch probes should be the same for all probes in a probe set. In detail, one first excludes any probe pair where $MM_i > 46000$ or where $|PM_i - MM_i| < \tau$ (where $\tau = 0.015$ by default), and then computes

$$R_j = \frac{PM_i - MM_i}{PM_i + MM_i},$$

which is known as the discrimination score. Suppose we have m discrimination scores for the probe set under consideration. One then uses Wilcoxon's signed rank test to test the hypothesis that the absolute value of the median discrimination score less τ is equal to zero. One can conduct this test by first finding the rank of the absolute value of each discrimination score, denoted r_j. Then one finds $r_j^* = \text{sgn}(R_j)r_j$, where $\text{sgn}(x) = 1$ if $x > 0$ and $\text{sgn}(x) = -1$ if $x < 0$, and computes the sum of the positive r_j^* to obtain a test statistic. One computes a p-value for the test statistic by comparing how extreme the observed value of this statistic is relative to the sampling distribution of the statistic assuming all 2^m possible values of the test statistic are equally likely (as they are under the null hypothesis). A transcript is said to be present if the p-value is less than 0.04, is absent if the p-value exceeds 0.06 and is called marginal otherwise.

While this test does provide a nonparametric method for evaluating if a transcript is present in a sample, it assumes the discrimination scores for probes within a probe set are independent and identically distributed, and this may not be true (since the cRNA probes for some probes within a probe set overlap for some arrays). Another weakness of this method for ascertaining if a transcript is present in a sample is that there are frequently replicate arrays in an experiment and these replicates aren't all used in the calculation of the presence calls. In fact, it is common for the presence calls to differ from replicate to replicate and this has led researchers to develop *ad hoc* criteria for combining presence calls across replicates (such as a transcript being called present if it is called present for a certain proportion of the replicates). Such *ad hoc* criteria make it very difficult to determine the actual significance level used in a test of if the gene is absent. In addition, failure to reject a null hypothesis is not equivalent to finding evidence for the alternative, hence the absent calls are not on the same footing as a presense call since the test may have had low power. Another concern is that many genes have multiple probe sets on these arrays and this information is not all used (in fact, it is possible for probe sets designed to measure the same gene give different results). Finally, one expects to falsely reject 5% of a set of nulls even if they are all true, and here we are conducting on the order of 10,000 hypothesis tests since microarrays frequently have this many spots.

12.4.2 Model based expression index

If one has replicate arrays for each experimental condition, then there are a number of ways one can improve estimation of gene expression levels by combining information across arrays. One of the first techniques that tried to do this was developed by Li and Wong and is referred to as the model based expression index (MBEI) (see Schadt et al., 2000; Li and Wong, 2001a and 2001b). The fundamental insight that motivates this method is that probes within a probe set differ in regard to their mean level, and if one has replicates arrays then one can determine these probe specific effects. Using this information, one can construct a more efficient estimate of the level of gene expression since our model captures an important feature of the data. When the method was proposed, the authors used the estimates from the model in conjunction with invariant set normalization (see Section 12.5.5 for further details). In addition, methods for identifying outliers were proposed.

The model used to provide the estimates assumes the existence of probe effects in addition to gene effects and estimates these parameters using least squares (which we can interpret as maximum likelihood with normally distributed errors as discussed in Section 2.3). We suppose that θ is the level of gene expression for some gene and that each probe has an effect specific to that probe denoted, ϕ_j. If we use the notation y_{ij} to represent the measured difference in intensity between the perfect match and the mismatch probes for probe pair j on array i (suppressing the dependence of these values on the gene index), we then suppose that $\mathrm{E}y_{ij} = \theta\phi_j$ and $\mathrm{Var}y_{ij} = \sigma^2$ (or alternatively $y_{ij} \sim \mathrm{N}(\theta\phi_j, \sigma^2)$) and that all y_{ij} are independent. If one constrains the ϕ_j so that $\sum_j \phi_j = m$ (where m is the number of probe pairs for this gene), then one can use least squares (or maximum likelihood) to estimate the parameters in the model. The need from the constraint arises due to the parameterization: if we decrease all ϕ_j by a factor of c and increase the value of θ by this same factor then the product appearing in the mean of the distribution of the probe differences is unchanged. This implies there are an infinite set of parameter values that give rise to the same value of the likelihood, so the model is unidentified. Hence the relative sizes of the ϕ_j parameters indicate the relative impact of the probes on the estimation of the level of gene expression.

12.4.3 Robust multi-array average

The robust multi-array average (RMA) is a widely used algorithm that incorporates many of the ideas that have already been discussed (see, for example, Irizarry et al., 2003). The entire algorithm entails choices regarding normalization, background correction and probe level summaries for Affymetrix microarrays. One can use the probe level summary in conjunction with other normalization and background correction procedures. The algorithm uses quantile normalization (see Section 12.5.2) and a background correction algorithm based on a parametric model. Only the perfect match data are used, and median polish is used to estimate array and probe specific effects.

One first uses quantile normalization on the perfect match probe intensities. Since the mismatch probe values are not used, RMA employs a background subtraction strategy in an attempt to account for non-specific binding. We suppose that each array has a common mean background level, so that if ϵ_{ijk} represents the background effect and η_{ijk} represents the signal measured by the perfect match probe, here denoted y_{ijk} (here k indexes gene while i indexes arrays and j indexes probe), then we have $y_{ijk} = \eta_{ijk} + \epsilon_{ijk}$. If one then assumes that the ϵ_{ijk} are all normally distributed with mean β_i and variance σ_i^2, while the η_{ijk} are all exponentially distributed with rate parameter λ_i, and all of these random variables are independent, then there is a closed form solution to $\mathrm{E}(\eta_{ijk}|y_{ijk})$, and this is used as the estimate of the background corrected estimate of the intensity of this probe. To obtain this solution, note that by Bayes theorem we have

$$p(\eta_{ijk}|y_{ijk}) \quad \propto \quad p(y_{ijk}|\eta_{ijk})p(\eta_{ijk}). \tag{12.1}$$

Next note that conditional on η_{ijn}, y_{ijn} is just normally distributed since it is the sum of a random variable we are conditioning on, η_{ijn} (which means we treat it as a constant) and a normally distributed random variable (i.e. ϵ_{ijn}). Thus $p(y_{ijn}|\eta_{ijn})$ will be a normal density with mean $\eta_{ijn} + \beta_i$ and variance σ_i^2. Then using the usual expressions for the densities of a normal with mean $\eta_{ijn} + \beta_i$ and variance σ_i^2 and an exponential random variable (with mean $\frac{1}{\lambda_i}$) in conjunction with equation 12.1, we find that

$$p(\eta_{ijk}|y_{ijk}) \quad \propto \quad \exp\left\{-\frac{1}{2\sigma_i^2}(y_{ijk} - \beta_i - \eta_{ijk})^2 - \lambda_i\eta_{ijk}\right\}$$

$$\propto \quad \exp\left\{-\frac{1}{2\sigma_i^2}(y_{ijn} - \beta_i - \sigma_i^2\lambda_i - \eta_{ijn})^2\right\}$$

so that $p(\eta_{ijk}|y_{ijn})$ is a normal distribution with mean $y_{ijk} - \beta_i - \sigma_i^2\lambda_i$ so that

$$\mathrm{E}[\eta_{ijk}|y_{ijk}] = y_{ijk} - \beta_i - \sigma_i^2\lambda_i.$$

Provided we have estimates of all parameters appearing in this expression we can then obtain explicit expressions for these conditional expectations. While one can derive a density for the marginal distribution of the y_{ijn} as it depends on the parameters β_i, σ_i^2 and λ_i and thereby obtain the MLEs of these 3 parameters, there is no explicit solution for these MLEs. However, since there are many data ponts that are informative about these 3 parameters (since they govern the marginal distribution of all of the perfect match probes on an array) one can just use a simple estimate, such as the method of moments estimate of $\beta_i + \sigma_i^2 \lambda_i$. As discussed further in exercise 5, if we let M_1 denote the first moment of y_{ijk} and M_m^C denote the mth central moment of y_{ijk} (i.e. $\mathrm{E}(y_{ijk} - \mathrm{E}y_{ijk})^m$) and if we denote the sample estimates of these moments with the same symbol but with a hat, then we find that the desired method of moments estimate is

$$\hat{M}_1 + \left(\frac{2}{\hat{M}_3^C}\right)^{\frac{1}{3}} \hat{M}_2^C - 2^{\frac{2}{3}}(\hat{M}_3^C)^{\frac{1}{3}},$$

hence we subtract this quantity from PM_{ijk} to get the desired conditional expectation.

The normalized, background corrected and log transformed perfect match intensities are then used along with median polish to estimate an array specific and probe specific additive effect for each gene. Median polish is an algorithm that supposes the median value of some variable is determined by 2 additive factors and additive error. Here the factors are the array specific effect (which is frequently a subject level effect) and a probe specific effect. The algorithm proceeds by constructing a 2 way table for each gene where the level of gene expression (after normalization, background correction and taking the logarithm) is given as it depends on the array and the probe. One then computes the median in each row, then subtracts this median from each of the rows. One saves the median for each of the rows for further modification. After this, one finds the median in each column and subtracts this value from each of the columns and stores these medians. Following the operation on the columns, one returns to the rows, computing the medians and subtracting those from each of the rows, including the median that was previously computed for that row. This process cycles back and forth between the rows and columns until the median in each row and column is zero. The process is illustrated in Table 12.1. Here we suppose that we have data for 3 arrays and the probe set for this particular gene has 5 probes (such low values for both are just for illustration). We record the changes in the median as the algorithm progresses along the margin for the particular array and probe. Here, the level of gene expression for this gene for each array is just the array level effect that appears in the last column in the final table.

One unusual feature of this model is that statisticians usually model effects of interest as differing between patient subgroups, not between different subjects (as the array specific effect here is usually subject specific as one has 1 array for each subject). In practice, one would first use RMA to get a gene level summary for each array, then average results from arrays in the same patient subcategory if one wanted to compare different patient populations. Based on observations regarding the probe sequence's impact on the hybridization intensity (Zhang, Miles and Aldape, 2003), this approach has been extended to include a model of how the nucleotide sequence of the probe impacts the hybridization intensity (Wu and Irizarry, 2005), and the resulting algorithm is called GC-RMA since the algorithm takes account of the GC content of the probe and target (since this impacts the hybridization energy).

Table 12.1 *An example of the median polish algorithm as applied to a probe set with 5 probes for 3 arrays. The algorithm has converged after 4 steps and the array level summaries are shown on the right margin of the last table.*

Original Data						Result of first step					
6.1	8.4	3.6	7.2	11.4	0.0	-1.1	1.2	-3.6	0.0	4.2	7.2
5.6	7.1	1.4	4.2	9.0	0.0	0.0	1.5	-4.2	-1.4	3.4	5.6
3.6	6.7	0.8	6.9	9.2	0.0	-3.1	0.0	-5.9	0.2	2.5	6.7
0.0	0.0	0.0	0.0	0.0		0.0	0.0	0.0	0.0	0.0	

Result of second step						Result of third step					
0.0	0.0	0.6	0.0	0.8	7.2	0.0	0.0	0.6	0.0	0.8	7.2
1.1	0.3	0.0	-1.4	0.0	5.6	1.1	0.3	0.0	-1.4	0.0	5.6
-2.0	-1.2	-1.7	0.2	-0.9	6.7	-0.8	0.0	-0.5	1.4	0.3	5.5
-1.1	1.2	-4.2	0.0	3.4		-1.1	1.2	-4.2	0.0	3.4	

Result of fourth step					
0.0	0.0	0.6	0.0	0.5	7.2
1.1	0.3	0.0	-1.4	-0.3	5.6
-0.8	0.0	-0.5	1.4	0.0	5.5
-1.1	1.2	-4.2	0.0	3.7	

12.5 Normalization

The goal of normalization is to reduce the effect of non-biological sources of variation on the measured level of gene expression. Typically it is assumed that these sources of variation operate at the level of the entire microarray, or a channel in the context of a spotted cDNA microarray. For example, suppose one of the 2 dyes used to label the samples is systematically integrated into the sample more efficiently than the other dye. If this was the case and there were no differences in the level of gene expression between the 2 samples, it would appear that the sample that is labeled with the dye that is more efficiently integrated into the target sample has a higher level of gene expression for every gene. Similarly, suppose that the hybridization reaction conditions were such that for some array, more of the target sample bound to the array. This could potentially happen if the target sample was allowed to be in contact with the array surface for a longer period of time. Then that array would appear to have a higher level of gene expression for all genes even though this is simply a consequence of differences in sample preparation. As we shall see, almost all of these corrections are motivated by assuming that only a small set of genes differ across samples. Tseng et al. (2001) provides a good introduction to many issues related to normalization.

Normalization is conducted after the basic quantification step and is usually applied at the level of an array, although there is some evidence that there are some within array normalizations that should be conducted. For example, in the context of spotted cDNA arrays, suppose that the print tips that spot the sequences onto the microarray differ in the ability of the tips to spot the cDNA onto the glass surface. Then any spot produced by a faulty tip would appear to have a lower intensity (see, for example, Yang et al., 2002). If one analyzes ratios of intensity then the differences due to print tips should cancel when one takes the ratio. Nonetheless, there are other potential sources of error that may lead to excess variation and can be remedied via normalization. Here we consider several popular normalization methods. It is generally believed that the normalization procedure one uses can have a substantial impact on the usefulness of the resulting set of gene expression level estimates (see Hoffman, Seidl and Dugas, 2002).

12.5.1 Global (or linear) normalization

A number of methods fall into this category: the unifying theme of these methods is that there is a multiplicative effect that is common across all spots on each array that must be accounted for prior to combining the results for multiple arrays. The most simple method in this category is to multiply all intensities on the array so that the mean (or median, or trimmed mean) level of expression across all transcripts on an array is constant from array to array(or constant across channels when more than 1 channel is used). This would account for overall differences between arrays in the conditions that impact hybridization efficiency for all transcripts. If we use y_i to represent the intensity of spot i, then this method corresponds to a model in which

$$\log(y_i) = \mu_i + \theta + \epsilon_i,$$

where ϵ is a zero mean error term, μ_i is the mean of the logarithm of the level of gene expression for spot i (averaging over all possible hybridizations) and θ is an array specific effect that we are trying to account for. Note also that we may suspect higher moments (such as the variance) are constant across arrays; this, too, would lead to a linear transformation of all probe intensities on the array using a common linear transformation for all transcripts. Finally note that these methods don't assume that some aspect of the distribution of gene expression is constant across arrays at the level of each gene. For example, if all genes are constant across 2 arrays except for 2 genes and one of these is higher on one array by the same amount that the other gene is lower, then the average across all transcripts can be the same even though the values for some genes differ. While only 2 genes differing across arrays by amounts that exactly cancel is not realistic, the usual argument in favor of this approach is that about equal numbers of genes will be expressed at higher levels as are expressed at lower levels and these groups roughly cancel. Clearly this won't always be the case, hence care should be taken when applying normalization methods.

Another less commonly used method for global normalization is to suppose that the level of expression for some gene (or set of genes) is constant across arrays. We refer to this as control gene normalization. This also leads to a linear transformation of the level of gene expression for each gene on the array. While analogous methods are frequently used in laboratory work, the 2 problems that one encounters in practice is selection of the spots that don't vary across arrays and the compounding of error that results from standardizing array results to one noisy observation (see exercise 6).

12.5.2 Spatially varying normalization

A slightly more sophisticated approach is to realize that there may be large scale effects on the array that vary across the surface. For example, suppose the array is darker the closer one is to a certain edge. This would suggest that a plane be fit to the array intensities, and we should then use residuals from these linear trends to quantify the level of gene expression. These more general formulations correspond to a model in which

$$\log(y_i) = \mu_i + \theta_i + \epsilon_i$$

where θ_i is constrained in some fashion. For example, if (z_{1i}, z_{2i}) are the Euclidean coordinates for probe i on the array, then setting

$$\theta_i = \beta_0 + \beta_1 z_{1i} + \beta_2 z_{2i}$$

would correspond to linear trends in intensity across the array surface. Moreover with either replicate probes for some transcripts or assumptions about the average level of gene expression across different arrays (for example, the level is constant) then parameter estimates can be obtained using least squares. Moreover, one can allow θ_i to vary non-linearly over the array surface in an arbitrary manner using methods from non-parametric regression.

12.5.3 Loess normalization

One of the first methods for normalization to gain widespread use for normalization of cDNA arrays is known as loess normalization. This method is similar to spatially varying normalization except no information about the location of the transcript on the array is used. Similarly though, if very few genes are differentially expressed, plotting the intensity for all genes for 2 arrays (or 2 channels) should show no pattern. Any pattern, such as nonlinear curvature, must be attributable to non-biological variation, and should thereby be corrected. Hence, to normalize 2 arrays, one plots the the difference in expression level for each gene against the average for that gene (this is just a rotation and rescaling of a plot of the 2 intensities from the 2 arrays/channels and is called an MA plot). One then uses the loess non-linear smoother to obtain a curve that is then used to normalize the 2 arrays: after applying the nonlinear transformation there should be no curvature in the MA plot.

In practice one must specify the span used in the loess smoothing algorithm. Given a value for this parameter, the loess algorithm uses a window of x values to estimate the value of $f(x)$, where this estimation is done using a robust average. This results in a robust local non-linear regression. The smaller the value of the span argument, the more variable the estimate of $f(x)$. For practical purposes, a value of around 0.3 is commonly recommended. Further information regarding this approach can be found in Yang et al. (2001), Yang et al. (2002), and Dudoit et al. (2002).

12.5.4 Quantile normalization

Quantile normalization (Bolstad et al., 2003) is an extension of the global methods introduced above. Those methods can be thought of as being motivated by the assumption that some marginal moments of the distribution of the gene expression measurements is the same across different arrays: for example the mean is constant across arrays. Quantile normalization can be thought of as being motivated by the assumption that all moments are the same across arrays, or equivalently, that the marginal cumulative distribution function of gene expression is constant across arrays (note that we don't assume that the individual gene expression levels are iid with this common distribution). In detail, let there be n arrays and each has p genes. Let $q_k = (q_{k1}, \ldots, q_{kn}), k = 1, \ldots, p$ be the vector of the kth quantiles for the n arrays and set $X = (q'_1, \ldots, q'_p)$. Next sort each column of X to get X_{sort}. Then assign the means across rows of X_{sort} to each element in the row to get X^*_{sort}. Finally, get $X_{\text{normalized}}$ by rearranging each column of X^*_{sort} to have the same ordering as X. This algorithm guarantees the marginal distribution of the gene expression levels across arrays is identical after normalization since all arrays will have the same values, but these values will be assigned to different spots on the different arrays.

12.5.5 Invariant set normalization

One commonly used normalization method for Affymetrix oligonucleotide arrays is invariant set normalization, introduced by Li and Wong (2001). In this approach, one first computes the proportion rank difference (PRD) (i.e. the absolute rank difference for some gene between two arrays divided by the number of genes on the array) for each gene using the perfect match probes. The invariant set of genes is then defined as those genes with PRD < 0.003 if the average intensity of the gene is small and PRD < 0.007 when the average is large. The normalization curve is then found by computing the running median in the scatterplot of probe intensities for two arrays. Aside from the use of rather ad hoc cutoffs for the PRD, when the invariant set is small, the running median curve may not perform well, and when there are many genes that differ across arrays, it's hard to find a large number of unchanged genes. Extensions of invariant set normalization have been developed within a Bayesian context that allow the user to specify prior probabilities that certain genes are in the invariant set, thereby avoiding the ad hoc cutoffs (Reilly, Wang and Rutherford, 2003) and will be discussed in section 13.5.

12.6 Exercises

1. Describe in detail how the signal value is computed from the y_i.

2. Show that it is possible that $\lambda < 1$ in the context of computing the signal value using the MAS 5.0 algorithm. Is this cause for concern? Explain.

3. Show that using the discrimination score is equivalent to using PM/MM in the presence call.

4. Give a normalization scheme that is appropriate if the variance is constant across arrays.

5. Show that the following estimate

$$\hat{M}_1 + \left(\frac{2}{\hat{M}_3^C} \right)^{\frac{1}{3}} \hat{M}_2^C - 2^{\frac{2}{3}} (\hat{M}_3^C)^{\frac{1}{3}},$$

is consistent for $\beta_i + \sigma_i^2 \lambda_i$ by computing the first moment and higher central moments of y_{ijk}.

6. Illustrate how one may compound error if one uses control gene normalization.

7. In a paper by Wang et al. (2007), the authors use Affymetrix microarrays to find genes in epithelial cells whose level of expression differs due to the level of Lim-only protein 4 (called LMO4). The probe level data is available through the gene expression omnibus (accession number GSE7382). Using the probe level data, compute MBEI and RMA for all genes on the arrays. Then use these gene level summaries to find genes that differ between the groups using a 2-sample t-test and a significance level of 0.01 (assuming equal variances across the groups). Are there more genes found to differ than you expect by chance?

CHAPTER 13

Detecting Differential Expression

13.1 Introduction

Given gene expression data from some set of experimental conditions, one of the first questions that the analysts wants to answer is which genes differ across conditions. At first glance, many of these questions can be dealt with most simply by conducting an analysis on a gene by gene basis using some spot level summary such as RMA. For the simplest sort of analysis, suppose we have a certain number of biological replicates from 2 distinct classes and want to address the question of which genes differ between the 2 classes. Anyone with some training in statistics would recognize this as the classical 2 sample problem and would know how to proceed. One would perhaps examine the distribution of the measurements within each class using a histogram to assess if the data depart strongly from normality and consider appropriate transformations of the data or perhaps even a non-parametric test (such as the Wilcoxon test). One would perhaps assess if the variances were unequal and perhaps correct for this by using Welch's modified 2 sample t-test, otherwise one could use the usual common variance 2 sample t-test. While there is some evidence that this is not optimal, such a procedure would be appropriate. As an example, consider Figure 13.1. This figure displays a histogram of the p-values one obtains if one uses a standard 2 sample t-test to test for differences in gene expression between a set of HIV negative patients and a set of HIV positive patients for all genes on an Affymetrix microarray. The tissue used in this application was taken from the lymph nodes of the patients, and we expect large differences between HIV negative and HIV positive patients in this organ because HIV attacks cells that are important for normal functioning of the immune system. If no genes are differentially expressed, then we expect this figure to look like a histogram of a uniform distribution, however that appears to not be the case since many genes have small p-values (for this example, over 12,000 genes have p-values that are less than 0.05).

While one can address many problems of interest using the standard statistical tools on gene level summaries, there are 2 important aspects of microarray studies that this fails to consider: the large number of genes and the usually relatively small number of replicates. This implies that due to the large number of genes, much is known about the sampling distribution of any statistic, and on the other hand, graphical tools at

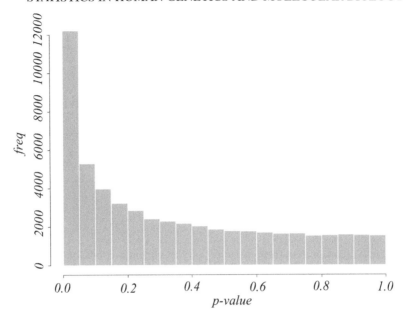

Figure 13.1 *A histogram of p-values that test for differences between 2 disease states using a standard 2 sample t-test for all 54,631 genes represented on an Affymetrix microarray.*

the level of the individual gene are not of much use. In addition, due to the small number of replicates, if we can do some sort of smoothing across genes (especially in the estimation of variances) then this is likely to be beneficial. Finally, gene level analysis is suboptimal when one notes that there is substantial correlation structure in the measured values of gene expression. While this last point is very difficult to take advantage of in the context of determining which genes are differentially expressed (due to the large number of genes) elucidating this correlation structure is an implicit goal of the next chapter and the basis for the final chapter.

13.2 Multiple testing and the false discovery rate

The problem of large scale multiple hypothesis testing has received a great deal of attention to those working on the analysis of microarrays (see, for example, Storey, 2002; Storey and Tibshirani, 2003). The problem stems from the fact that if one conducts p hypothesis tests and each uses the significance level α, then if all of the null hypotheses are true and all tests are independent

$$P(\text{at least 1 gene is found to be differentially expressed})$$

$$= 1 - P(\text{no genes are found to be differentially expressed})$$

$$= 1 - (1 - \alpha)^p$$

so if we use the conventional value of $\alpha = 0.05$ and we test 1,000 genes then the probability that we find at least 1 gene that is differentially expressed is $1 - 0.95^{1000} \approx 1 - 5.3 \times 10^{-23}$ when in fact none are differentially expressed. Thus we are virtually guaranteed to find at least 1 gene that differs between conditions even if no genes really do differ.

The relevance of this observation varies with the application. Suppose that one is attempting to identify genes whose level of expression differs between 2 patient populations. If the patients differ due to their status with respect to skin cancer and one is looking for differentially expressed genes in urine, it is not obvious that any genes will be differentially expressed. In contrast, if one was to look for genes that differ in their level of expression between cancerous skin cells (from the affected group) and healthy skin tissue samples (from the control group) then one would expect before even looking at the data that there will be differentially expressed genes (since the observed phenotypes differ and this is regulated by the repertoire of genes being expressed in the cell). While one could argue in the latter case that it is possible that the phenotypes differ between the cell populations for reasons other than gene expression (e.g. differences in the protein composition between the 2 cell populations) one is still more inclined to think that there are differentially expressed genes when one examines cells with different phenotypes.

In the case where one wants to see if there is evidence for any genes being differentially expressed, one would attempt to control the family-wide error rate for all tests, i.e. the probability of falsely rejecting any null hypothesis. Note that by a result called Bonferroni's inequality, if A_i are a set of events then

$$P(\cup_i A_i) \leq \sum_i P(A_i).$$

We can use this inequality to control the probability that we find any gene is differentially expressed when in fact no genes are differentially expressed. To do so, let A_i be the event that we reject hypothesis i. Let H_{0i} represent the null hypothesis of no differential expression for gene i where $i = 1, \ldots, p$, and we use the notation that $p_{H_{0i}}(A)$ is the probability of event A assuming H_{0i}. Let α^* be the family wide error rate for all tests. Here we assume that $P_{H_{0i}}(A_i) = P_{\cap_i H_{0i}}(A_i)$ so that the nulls are mutually uninformative. Then it follows from Bonferroni's inequality that

$$
\begin{aligned}
\alpha^* &= P_{\cap_i H_{0i}}(\cup_i A_i) \\
&\leq \sum_i P_{\cap_i H_{0i}}(A_i) \\
&\leq \sum_i P_{H_{0i}}(A_i).
\end{aligned}
$$

Since the p-value is the probability of observing a value for a statistic that is greater than what we have observed assuming that the null is true, we can control each probability on the right hand side by only rejecting null hypothesis i if $p_i < \alpha_i$ for some α_i. For example, if we set $\alpha_i = \alpha/p$ for all i then $\alpha^* \leq \alpha$ and so we can control the family-wide error rate for all tests at level α (this is known as Bonferroni's method for adjusting for multiple tests and use of the method is referred to as Bonferroni's correction). The main problem with this approach is that with many genes α_i will be very small, for example, if we have 10,000 genes and want $\alpha^* \leq 0.05$ then we need to set $\alpha_i = 5.0 \times 10^{-6}$ and so each test will have very little power. Alternatively, we could select a set of size p_1 genes of primary importance to study and use $\alpha_1^* = \frac{\alpha_1}{p_1}$ for these genes and use $\alpha_2^* = \frac{\alpha_2}{m - p_1}$ for the remaining genes where $\alpha_1 + \alpha_2 = \alpha$. This would at least provide greater power for the genes of primary interest. However, there are also a number of more fundamental issues related to controlling the family-wide error rate. For instance, if I present the results from more than 1 set of arrays, should I control my error rate across all sets of arrays? Should this control happen at the level of a manuscript? Perhaps journals should control the family-wide error rate for each issue? These considerations start to make control of the family-wide error rate problematic. Finally, note that the rejection of the null is just that there is at least 1 gene that is differentially expressed, not that any particular gene is differentially expressed. Nonetheless, control of the family-wide error rate was quite common in early publications, but perhaps that was a sensible approach since microarray technology was not as widely used (and supported) as it is today.

To avoid problems associated with controlling the family-wide error rate, researchers in genomics have increasingly used the false discovery rate (FDR) to generate a list of genes that are thought to be differentially expressed. This criterion leads to tests that are more powerful (thereby identifying more genes) and has a convenient interpretation. The FDR is the expected proportion of genes that do not differ between groups of those genes that are declared to differ. Thus if one has a list of 53 genes and the FDR is 10%, then one would expect that about 5 of the genes on that list do not really differ between groups. If we suppose that we will declare S genes as significant based on some criterion (such as p-values for gene level tests) and F of these are not differentially expressed, then FDR$= \mathrm{E}\left[\frac{F}{S}\right]$. Note that with positive probability $S = 0$ so that FDR is undefined. There is work to address this problem (the so called positive false discovery rate, see Storey, 2003), but we will just consider the FDR here.

To be more concrete, suppose there are p genes and we want to find which of these are differentially expressed between 2 groups by using a 2-sample t-test for each gene. This would then generate a set of p p-values. Frequently a histogram of these p-values will not look like a uniform distribution on $(0,1)$ (which it should if all nulls were true), but rather has many more small p-values than one would expect by chance as in Figure 13.1. The question then becomes how small does the p-value have to be so that we declare some gene to be differentially expressed. We assume that if one thinks some p-value, say p_i, is so small that we should deem that gene differentially expressed then we should also declare any gene with a p-value smaller than p_i to be differentially expressed. This is a reasonable assumption if one only has the p-values for the tests and knows nothing about the dependence of p-values across genes. Under this assumption, the problem is then just a matter of deciding what threshold to use on the set of p-values so that genes with p-values smaller than this threshold are declared differentially expressed. The idea behind the FDR is that

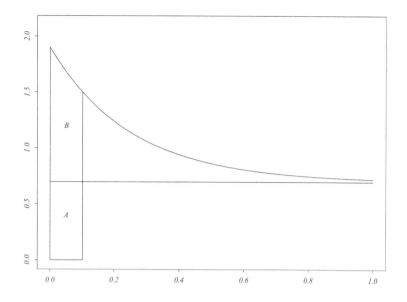

Figure 13.2 *A graphical representation of the FDR. If we reject the null for all genes with p-values less than 0.1 the FDR would be A/(A+B).*

given any threshold (denoted t for some $0 \leq t \leq 1$), we will attempt to estimate the FDR (which we will now denote $\mathrm{FDR}(t)$) associated with this threshold.

For large p, it transpires that $\mathrm{FDR}(t) \approx \frac{\mathrm{E}F(t)}{\mathrm{E}S(t)}$. A sensible estimate of $\mathrm{E}S(t)$ is just $S(t)$ which is the number of observed p-values that are less than or equal to t. For the numerator, we use the fact that if p_0 is the number of genes for which the null is true, then we expect $p_0 t$ of these to have p-values $\leq t$ since the distribution of these p-values under the null is uniform on $(0,1)$. Hence we need an estimate of p_0, or equivalently $\pi_0 = p_0/p$, the proportion of genes that are not differentially expressed between the groups. To address this we suppose that the distribution of p-values that we observe is a mixture of p-values from genes that are not differentially expressed (which has a uniform distribution on $(0,1)$) and some other distribution that places more mass on lower values which we denote $f_1(t)$. Hence the distribution of observed p-values has density

$$p(t) = (1 - \pi_0)f_1(t) + \pi_0,$$

and so

$$\lim_{t \to 1} p(t) \quad = \quad (1 - \pi_0)f_1(1) + \pi_0$$

$$\geq \quad \pi_0.$$

This idea is illustrated in Figure 13.2. Thus if we can estimate $\lim_{t \to 1} p(t)$ we would have a quantity that estimates a quantity greater than π_0 (although the estimate itself may not be greater than π_0), and so the estimate is conservative in that on average it will declare fewer genes as differentially expressed than is the case. Note that we don't expect $f_1(t)$ to be large near 1 since this is the density for p-values of genes that are differentially expressed, so the estimate of π_0 will not be too large.

While a number of methods have been proposed to estimate $\lim_{t \to 1} p(t)$, a simple approach is based on the idea that there is some value λ so that $\int_\lambda^1 f_1(t)\,dt \approx 0$. Using this we have

$$
\int_\lambda^1 p(t)\,dt \quad \approx \quad \int_\lambda^1 \pi_0
$$
$$
\approx \quad (1-\lambda)\pi_0,
$$

and since an unbiased estimate of $\int_\lambda^1 p(t)\,dt$ is provided by $\frac{1}{p}\sum_i 1_{\{p_i > \lambda\}}$ we have the following estimate of π_0,

$$
\hat{\pi}_0(\lambda) = \frac{\sum_i 1_{\{p_i > \lambda\}}}{p(1-\lambda)}.
$$

While one can simply look at a histogram of p-values and find a reasonable value for λ, one can also automate this computation in a number of ways. For example, we could compute $\hat{\pi}_0(\lambda)$ for some set of values λ_j where $j = 1, \ldots, 100$, $\lambda_1 = 0.01$ and $\lambda_{100} = 0.95$ then use a quadratic regression of $\hat{\pi}_0(\lambda_j)$ on λ_j to predict $\hat{\pi}_0(1)$ (a quadratic regression is necessary due to non-linearity in the plotted values). This is the approach of Storey and Tibshirani (2003). We then have the following estimate of

$$
\hat{\mathrm{FDR}}(t) = \frac{p\hat{\pi}_0(1)t}{S(t)},
$$

and by increasing t we find that more genes have p-values below the threshold and the estimated FDR rises.

Notice that as t increases more genes are declared significant, thus for each gene there is a minimal value of the estimated FDR so that that gene is in the list of genes that are declared to be differentially expressed. Hence, for each gene we can estimate the expected proportion of false discoveries if that gene is declared differentially expressed (it is simply the minimal FDR so that a gene is in the list of those declared to differ). These gene level estimates are called q-values. While the use of a FDR of 0.1 is common, the choice can be made after examining the results. For instance, by plotting the number of genes declared to differ by the q-values one can sometimes detect certain values for the q-value so that there is a large jump in the number of genes declared to be significant. This would indicate that there are some values where the FDR is nearly constant while the number of genes declared to differ jumps sharply (this can also be detected by examining the relationship between the number of genes declared to differ and the expected number of false discoveries).

13.3 Significance analysis for microarrays

There is a popular algorithm known as significance analysis of microarrays (SAM) (see Gosh Tusher, Tibsharani, and Chu, 2001; Efron et al., 2001) that incorporates several of the general methods for summarizing gene expression levels, such as non-parametric inference, smoothing of the variance estimates (also known as regularization of the variance estimates; see, for example, Baldi and Long, 2001) and the use of the false discovery rate. The first statistical presentation of these ideas appeared in a paper which analyzed data from Affymetrix arrays, so we will present the algorithm in that context (extending the basic idea to other settings, such as cDNA microarray data, is straightforward. That particular experiment involved 8 microarrays and investigated the effect of radiation (present at 2 levels: unirradiated or irradiated) on a set of cells. There were 2 cell types used in the experiments and 2 replicates of each array condition. If we refer to the cell types as types 1 and 2 and the replicates with the letters A and B, then we can refer to each of the arrays in the following notation: U1A, U1B, I1A, I1B, U2A, U2B, I2A, I2B (so for instance, U1A is one of the 2 replicated microarrays that used cell type 1 and were unirradiated). Below we will use the notation $U1A_i$ to refer to the gene level summary of gene i on the array with those conditions. One can think of the process of going from the probe level data (there are about 15-20 probe pairs for each gene) to a gene level summary on each array, then to a gene level summary that summarizes over all arrays. The final task is then to determine appropriate methods for statistical inference.

13.3.1 Gene level summaries

To summarize the probe level data for a probe set, if we let (PM_{ijk}, MM_{ijk}) be the values of the intensity measured for gene i, probe pair j for $j = 1, \ldots, J_i$ on array k, one first computes

$$M_{ik} = \frac{1}{J_i} \sum_{j=1}^{J_i} (\log PM_{ijk} - \frac{1}{2} \log MM_{ijk}).$$

Here M_{ik} summarizes the level of gene expression for this gene on array k, and is the average of the logarithm of the probe pair intensities where the mismatch probes are downweighted (the choice of the value of 0.5 will be discussed below). By using the log the effect of outliers is reduced, and by downweighting the mismatch probes one can potentially obtain an estimate that is less variable than equally weighting them.

Since the primary goal is to detect differences in gene expression that result from treatment with radiation, the gene level summaries examine differences in the level of gene expression between samples that are matched for all factors except irradiation status. Hence for each gene i we obtain the 4 values D_{ij} where $D_{i1} = I1A_i - U1A_i$, $D_{i2} = I1B_i - U1B_i$, $D_{i3} = I2A_i - U2A_i$, $D_{i4} = I2B_i - U2B_i$. Then for each gene one computes the mean of these 4 values, $\overline{D}_i$, and the standard deviation of these 4 values, s_i. The gene level summary for gene i is then given by

$$z_i = \frac{\overline{D}_i}{a_0 + s_i},$$

where a_0 is some percentile of the s_i (the 90th was used in the original paper). The choice of the use of the 90th percentile is dealt with in the same way that the choice of $\frac{1}{2}$ is determined in the gene level summary for each array and will be discussed below. Note that from the perspective of ranking the effect size $|z_i|$, for each gene, if we take a_0 to be large then the rank of the $|z_i|$ will be determined by $|\overline{D}_i|$, thus the ranks are determined by the differences in the 2 array condition (completely ignoring the standard deviations of the differences), which is equivalent to sorting the genes by fold change. On the other hand, if we set $a_0 = 0$ then the relative effect of the individual gene is determined by the paired t-test statistic (since this latter statistic is just $\sqrt{4z_i}$). By allowing intermediate values that are selected by the data this method has the potential to outperform either of these 2 approaches (for ranking genes) since they are both special cases.

13.3.2 Nonparametric inference

The gene level summaries provide a way of ranking the genes in terms of their relative effect sizes, but this doesn't indicate what critical value should be used for these test statistics (i.e. what value of $|z_i|$ is large enough for us to declare that some gene is differentially expressed between 2 samples). This will be addressed by examination of the marginal distribution of the entire set of the z_i when there is no difference between the 2 samples. Let π_1 be the probability that some gene is affected by radiation and let $f_1(z)$ be the distribution of the z_i values under this condition. If we define π_0 to the probability that a gene is not affected and $f_0(z)$ to the density of the z_i for these genes, then if $f(z)$ is the density of the z_i values averaging over if genes are affected by radiation, we can write

$$f(z) \quad = \quad \pi_0 f_0(z) + \pi_1 f_1(z). \tag{13.1}$$

Next note that is we use $\pi_1(z)$ to represent the probability that a gene is differentially expressed (DE) conditional on its measured value of z (so $\pi_1(z) = P(\text{DE}|z)$), then by Bayes theorem

$$\pi_1(z) \quad = \quad \frac{P(z|\text{DE})P(\text{DE})}{f(z)}$$

$$= \quad \frac{f_1(z)\pi_1}{f(z)}$$

but then by equation 13.1,

$$\pi_1(z) \quad = \quad \frac{f(z) - \pi_0 f_0(z)}{f(z)}$$

$$= \quad 1 - \frac{\pi_0 f_0(z)}{f(z)}.$$

Note that the probability of differential expression conditional on z just depends on $f(z)$, $f_0(z)$ and π_0. Any microarray data set will have information about $f(z)$, and by examining sets of z_i after certain permutations one can obtain information about $f_0(z)$(this is discussed more fully below), but the determination of π_0 is not so obvious (the method for obtaining an estimate of π_0 from Section 13.2 can't be used here since we don't have a set of p-values). Nonetheless, given that $\pi_1(z) \geq 0$ we find that

$$1 - \frac{\pi_0 f_0(z)}{f(z)} \geq 0$$

hence we can conclude that

$$\pi_0 \leq \min_z \frac{f(z)}{f_0(z)},$$

and so we can get an upper bound on π_0 by using this inequality, so that

$$\hat{\pi}_0 = \min_z \frac{f(z)}{f_0(z)}.$$

Thus we estimate π_0 with our estimated value of $\min_z \frac{f(z)}{f_0(z)}$ and use this in our expressions for $\pi_1(z)$.

To estimate $f_0(z)$ we can create random realizations of the set of the z_i under the assumption of no differences in gene expression due to the experimental treatment. This can be obtained by computing d_{ij} where $d_{i1} = \text{U1B}_i - \text{U1A}_i$, $d_{i2} = \text{I1B}_i - \text{I1A}_i$, $d_{i3} = \text{U2B}_i - \text{U2A}_i$, $d_{i4} = \text{I2B}_i - \text{I2A}_i$, since each of these differences is just due to differences between replicates. Note that we can also switch the signs of each of the d_{ij} since the distribution should be symmetric. We can then generate an entire vector of z_i's by randomly changing the signs of the d_{ij} and computing z_i for each gene independently. Since there are many genes there are many distinct vectors of the z_i.

We generate artificial data sets to estimate the ratio of $f_0(z)$ to $f(z)$ and use logistic regression to relate these values to $\pi_1(z)$. Logistic regression is an extension of linear regression in which the outcome variable is binomial (instead of normal) and the binomial success probability depends on a set of predictor variables. If x_i is the vector of predictor variables associated with the binary outcome y_i and β is a vector of regression coefficients that give the effect of each element of x_i on the probability that $y_i = 1$ then logistic regression supposes that

$$y_i \sim \text{Ber}(n_i, \theta_i)$$

where

$$\log\left(\frac{\theta_i}{1 - \theta_i}\right) = x_i' \beta.$$

The MLE of β is obtained by numerically maximizing the likelihood as it depends on β.

Here, logistic regression is used as follows. First generate 20 vectors of z_i from its null distribution (20 was selected based on experience with the algorithm). We then create an "artificial data set" where the values of the explanatory variable are given by either the $20p$ values of z_i obtained from the null distribution and the p observed values of the z_i (based on the differences in expression due to radiation of the cells). We then associate with each $|z_i|$ either a 0 (if that z_i is from the null distribution) or a 1 (if that value is from the observed differences due to radiation). Denote these binary variables as y_i for $i = 1, \ldots, 21p$. Then note that by Bayes theorem we have

$$
\begin{aligned}
P(y_i = 1 | z) &= \frac{p(z|y_i = 1)p(y_i = 1)}{p(z|y_i = 1)p(y_i = 1) + p(z|y_i = 0)p(y_i = 0)} \\
&= \frac{f(z)\frac{1}{21}}{f(z)\frac{1}{21} + f_0(z)\frac{20}{21}}.
\end{aligned}
$$

Next, note that we can estimate $P(y_i||z|)$ using logistic regression where the explanatory variable is the absolute value of a gene level summary for some gene. In Efron et al. (2001), the authors propose to use a set of piecewise quadratic functions of z that are constrained to be continuous. Such functions of z are known as cubic splines and the authors propose to use cubic splines over the intervals defined by the tenth and ninetieth quantiles of the set of z_i that are observed. While this is more flexible than just using the absolute value of z, it is more complex and both approaches allow for $\pi_1(z)$ to increase as z moves away from zero.

If we then denote the corresponding estimate $\hat{\pi}(z)$ then we can get an estimate of $\pi_1(z)$ by using

$$
\hat{\pi}_1(z) = 1 - \hat{\pi}_0 \frac{1 - \hat{\pi}(z)}{20\hat{\pi}(z)}.
$$

This last equation provides a way for estimating the probability of differential expression based on a value we obtain from the reduction of the data to the gene level summary.

13.3.3 The role of the data reduction

There are a number of inputs to the probe set reduction and gene reduction that are data dependent in this algorithm, namely the use of $\frac{1}{2}$ in the probe set reduction and the selected value of a_0 in the gene level summary. Other aspects of the algorithm could in practice be selected using the same ideas used to select the values for these parameters. An examination of the above algorithm suggests a central role to the value of the ratio $\frac{f_0(z)}{f(z)}$. This ratio is just a function of the method used for data reduction. In particular, a good data reduction will lead to this ratio departing from 1 for large values of z. A very bad data reduction would lead to finding that this ratio is 1 for all z (thereby indicating no differences between the value of the statistic used for the data reduction under the null hypothesis and under the alternative). Hence given some input to the data reduction (like the value of a_0 used) one can examine this ratio as it depends on this input parameter. We then select the value of the input parameter to make our estimate of the ratio depart from 1 for large values of $|z|$.

13.3.4 Local false discovery rate

Recall that a basic assumption for estimating the FDR was that one has a set of p-values and if one declares some p-value small enough to say there is a difference, one must also declare there is a difference for all genes with smaller p-values. In the use of the local FDR, one assumes all genes with similar test statistics (i.e. z_i) to one we declare to differ must also be declared to differ. It also transpires that the use of this procedure can be used to control the local false discovery rate. In fact the estimated probability of a gene being not differentially expressed is approximately equal to the local false discovery rate. To see this, suppose there are a total of p genes and m genes have values of z_i that fall in some interval C. Suppose that the 20 draws from the null distribution result in m_1 genes that have z_i that fall in this interval, or $m_1/20$ per set. So if no genes differed between conditions we would expect $m_1/20$ genes to fall in this interval. If $\hat{\pi}_0 < 1$ then we think that some genes must differ between conditions, and we then expect $\hat{\pi}_0 m_1/20$ genes would have values of z_i that fall in this interval using the real values of the z_i that are really unaffected. Hence the estimated local false discovery rate is $\frac{\hat{\pi}_0 m_1/20}{m}$ since this is the proportion of genes that have values of the test statistic that lie in this interval that are probably unaffected. But note that for some small interval, we can estimate $f(z)$ using a histogram estimate which estimates the density by the number of items falling in that interval relative to the total number of items, i.e. $\hat{f}(z)$ is given by the proportion of genes that fall in the interval, so that the estimated local false discovery rate is just $\hat{\pi}_0 \frac{(m_1/20)/p}{m/p}$ as claimed.

13.4 Model based empirical Bayes approach

Bayesian approaches to the problem of detecting differential expression have also been developed (for example, Ibrahim, Chen and Gray, 2002; Bernardo, Bayarri and Berger, 2002; Reilly, Wang and Rutherford, 2003). Most of these methods borrow strength across genes to construct more efficient estimates (often at the level of the gene variance, as in other uses of regularization) and often use gene level indicator variables that indicate if a gene is differentially expressed. Here we discuss one method (Newton et al., 2001) that has been developed in the context of the analysis of cDNA microarrays. This method uses the ratio of the intensity in the 2 channels to quantify relative gene expression. In addition it uses empirical Bayes methods to construct estimates of the level of gene expression.

This method is based on the observation that while the intensity in the 2 channels in a cDNA array varies from gene to gene and between channels, many normalization methods results in a dataset in which the coefficient of variation (i.e. the ratio of the standard deviation to the mean) is largely constant. Hence the model should have constant coefficient of variation across genes. To this end, let g denote the intensity in one of the channels and r the intensity in the other channel for some gene. If we suppose that

$$g|\alpha, \theta_g \sim \mathrm{Gam}(\alpha, \theta_g)$$

and

$$r|\alpha, \theta_r \sim \text{Gam}(\alpha, \theta_r)$$

where the parameter α is constant across all genes and θ_g and θ_r are gene specific (although here we are suppressing this dependence). To see that this leads to a constant coefficient of variation note that

$$\text{E}[g|\alpha, \theta_g] = \frac{\alpha}{\theta_g}$$

and

$$\text{Var}[g|\alpha, \theta_g] = \frac{\alpha}{\theta_g^2}$$

so that the coefficient of variation is $\frac{1}{\sqrt{\alpha}}$ for all genes and for both channels. In this way, the model pools information from all the genes much like the SAM algorithm does when estimating the value in the denominator of the gene level summaries.

Next, if we specify independent conjugate priors for θ_g and θ_r (i.e. both of these parameters have $\text{Gam}(\alpha_0, \nu)$ priors for all genes), then (conditional on α, α_0 and ν) we find the posterior distribution of θ_g is $\text{Gam}(\alpha + \alpha_0, g + \nu)$ and the posterior distribution of θ_r is $\text{Gam}(\alpha + \alpha_0, r + \nu)$, and these 2 distributions are independent. As researchers are frequently interested in the fold change, here we will examine the posterior distribution of the ratio of gene expression across the 2 channels. To this end, we set $\rho = \theta_g/\theta_r$, then we find that the posterior density of ρ conditional on $\eta = (\alpha, \alpha_0, \nu)$ is proportional to

$$\rho^{-(\alpha+\alpha_0+1)}\left[\frac{1}{\rho} + \frac{g+\nu}{r+\nu}\right]^{-2(\alpha+\alpha_0)}$$

since

$$
\begin{aligned}
\frac{d}{dt}P\left(\frac{\theta_g}{\theta_r} \leq t|g, r, \eta\right) &= \frac{d}{dt}\int_0^\infty \int_0^{t\theta_r} p(\theta_g, \theta_r|g, r, \eta)\,d\theta_g\,d\theta_r \\
&= \int_0^\infty \theta_r p(\theta_r t|g, r, \eta)p(\theta_r|g, r, \eta)\,d\theta_r \\
&\propto \int_0^\infty \theta_r^{\alpha+\alpha_0}(\theta_r t)^{\alpha+\alpha_0-1}e^{-\theta_r(r+\nu)-\theta_r t(g+\nu)}\,d\theta_r \\
&\propto t^{\alpha+\alpha_0-1}[(r+\nu) + t(g+\nu)]^{-2(\alpha+\alpha_0)} \\
&\propto t^{-(\alpha+\alpha_0+1)}\left[\frac{1}{t} + \frac{g+\nu}{r+\nu}\right]^{-2(\alpha+\alpha_0)}.
\end{aligned}
$$

The posterior mode of ρ is then approximately $\frac{r+\nu}{g+\nu}$. Note that this mode has the properties of a shrinkage estimate: if ν is small the posterior mode is just determined by the ratio of r to g whereas if ν is large the posterior mode will be near 1. Thus depending on the value of ν we find that the posterior mode will be the ratio of the measured values "shrunk" towards the central value of 1. This is a result of assuming that the coefficient of variation is common across all genes.

While this approach provides us with a simple summary of the relative levels of gene expression in the 2 channels, there are still 2 major shortcomings. First, we still must determine a value for ν. In addition, since the point estimate of ρ derived above doesn't indicate if the estimate is sufficiently far from zero to conclude that the gene is differentially expressed, we would like some numerical summary indicating the probability that a given gene is differentially expressed. The first issue will be dealt with by employing methods from empirical Bayes analysis and the second via a common model expansion.

To determine an appropriate value of ν we derive the marginal posterior distribution of η and use an empirical Bayes approach in which we maximize this marginal posterior to find $\hat{\eta}$ and use $\hat{\nu}$ in our expressions for Bayes estimates of relative gene expression. Using the same conjugate prior for θ_g and θ_r as above (formally, these are conditional on η and we use a flat prior on η), we find that we can express the posterior distribution of θ_g, θ_r and η as

$$p(\theta_g, \theta_r, \eta | g, r) \quad \propto \quad \frac{\theta_g^\alpha}{\Gamma(\alpha)} g^{\alpha-1} e^{-g\theta_g} \frac{\theta_r^\alpha}{\Gamma(\alpha)} r^{\alpha-1} e^{-r\theta_r} \frac{\nu^{\alpha_0}}{\Gamma(\alpha_0)} \theta_g^{\alpha_0-1} e^{-\nu\theta_g}$$
$$\times \frac{\nu^{\alpha_0}}{\Gamma(\alpha_0)} \theta_r^{\alpha_0-1} e^{-\nu\theta_r}.$$

Hence if we integrate θ_g and θ_r out of the posterior we obtain an expression for the marginal posterior of η:

$$p_A(\eta | g, r) \quad \propto \quad \frac{(rg)^{\alpha-1} \nu^{2\alpha_0}}{\Gamma(\alpha)^2 \Gamma(\alpha_0)^2} \frac{\Gamma(\alpha + \alpha_0)^2}{[(r + \nu)(g + \nu)]^{\alpha+\alpha_0}}. \tag{13.2}$$

For the model discussed above, since we typically have thousands of genes on a microarray, the parameters α, α_0 and ν govern the marginal distribution of gene expression (i.e. the distribution of gene expression ratios averaging over all genes on the array), hence for a given data set, there is a great deal of information on likely values for these parameters. This suggests that we can approximate the posterior distribution of the level of gene expression (i.e. θ_g and θ_r) by finding the posterior mode of η and then substituting this mode into our expression for the posterior distribution of θ_g and θ_r. Note that the previous derivation of the marginal likelihood of α, α_0 and ν considered only a single gene: since all of the genes contribute to the posterior in a symmetrical manner, we find the marginal posterior for these parameters by summing over all genes, hence, if there are p genes, the log posterior of η is

$$\log p(\nu | g_1, r_1, \ldots, g_p, r_p) = \sum_{i=1}^{p} \log p_A(\eta | g_i, r_i).$$

We can then optimize this expression to obtain an estimate of ν which can then be substituted into our expression for the approximate posterior mode of ρ for all genes on the array, thereby providing an estimate of the relative level of gene expression for all genes on the array.

Finally, we can obtain the posterior probability that a gene is differentially expressed by introducing a parameter for each gene that is zero or one depending on if the gene is differentially expressed. If we then examine the posterior distribution of these indicator variables we can determine the posterior probability that a gene is differentially expressed. For each gene, let $\xi_i = 1$ if gene i is differentially expressed and 0 otherwise. We suppose that these ξ_i are all iid Bernoulli random variables with success probability π, and we use the notation $\xi = (\xi_1, \ldots, \xi_p)$. We then specify a prior for π (such as $\text{Beta}(c, c)$ with c set to a small value like 2). The goal now is to determine the posterior mean of ξ_i since this posterior mean is the posterior probability that $\xi_i = 1$, which corresponds to the posterior probability that gene i is differentially expressed. To do this, we will use the EM algorithm treating the ξ_i as missing data and $\gamma = (\eta, \pi)$ as the parameters in the model that are maximized in the M-step.

To use the EM algorithm, we first write down an expression for the complete data log-posterior. Here this is

$$p(\xi, \gamma | g_1, r_1 \ldots, g_p, r_p) \quad \propto \quad p(g_1, r_1 \ldots g_p, r_p | \xi, \gamma) \prod_{i=1}^{p} p(\xi_i | \pi)$$

$$\propto \quad \prod_{i=1}^{p} p(g_i, r_i | \xi_i, \eta) \prod_{i=1}^{p} \pi^{\xi_i} (1 - \pi)^{1 - \xi_i}. \quad (13.3)$$

Next note that

$$p(g_i, r_i | \xi_i, \eta) \quad = \quad \int \int p(\theta_g, \theta_r, g_i, r_i | \xi_i, \eta) \, d\theta_g \, d\theta_r$$

$$= \quad \int \int p(g_i, r_i | \theta_g, \theta_r, \xi_i, \eta) p(\theta_g, \theta_r | \xi_i, \eta) \, d\theta_g \, d\theta_r$$

and note that we already performed this integration in the case where $\xi_i = 1$ when we derived equation 13.2. Hence we will have an expression for the complete data posterior if we perform this integral for the case where $\xi_i = 0$, that is $\theta_g = \theta_r = \theta$. The reader is asked to find this expression in exercise 4, here we will denote the result as $p_0(\eta | g_i, r_i)$ which is given by the expression

$$p_0(\eta | g_i, r_i) = \frac{(g_i r_i)^{a-1} \nu^{a_0}}{(r_i + g_i + \nu)^{2a + a_0}} \frac{\Gamma(2a + a_0)}{\Gamma(a)^2 \Gamma(a_0)}.$$

Hence we find that $p(g_i, r_i | \xi_i, \eta)$ in equation 13.3 is $p_0(\eta | g, r)$ when $\xi = 0$ and $p_A(\eta | g, r)$ when $\xi = 1$, hence have

$$p(g_i, r_i | \xi_i, \eta) = p_A(\eta | g_i, r_i)^{\xi_i} p_0(\eta | g_i, r_i)^{1 - \xi_i},$$

and so

$$p(\xi, \gamma | g_1, r_1 \ldots g_p, r_p) \propto \prod_{i=1}^{p} p_A(\eta | g_i, r_i)^{\xi_i} p_0(\eta | g_i, r_i)^{1 - \xi_i} \pi^{\xi_i} (1 - \pi)^{1 - \xi_i}.$$

We can therefore write the complete data log-posterior in the form

$$p(\xi, \gamma | g_1, r_1 \ldots g_p, r_p) \;=\; \sum_{i=1}^{p} \Big[\xi_i \log p_A(\eta | g_i, r_i) + (1 - \xi_i) \log p_0(\eta | g_i, r_i) \\ + \; \xi_i \log \pi + (1 - \xi_i) \log(1 - \pi) \Big].$$

Since the ξ_i are missing, we first compute the expected value of these conditional on a starting value for (η, π). Now

$$\mathrm{E}[\xi_i | g_i, r_i, \eta, \pi] = P(\xi_i = 1 | g_i, r_i, \eta, \pi),$$

but since

$$P(\xi_i = 1 | g_i, r_i, \eta, \pi) \propto p(g_i, r_i, \eta, \pi | \xi_i = 1) p(\xi_i = 1),$$

we find that

$$\mathrm{E}[\xi_i | g_i, r_i, \eta, \pi] = \frac{\pi p_A(\eta | g_i, r_i)}{\pi p_A(\eta | g_i, r_i) + (1 - \pi) p_0(\eta | g_i, r_i)}.$$

So the E-step consists of computing all these expectations given previous values of (η, π). One then substitutes all of the expectations into the complete-data log-likelihood and maximizes the resulting expression using some optimization algorithm. When the algorithm converges, the resulting set of $\mathrm{E}[\xi_i | g_i, r_i, \eta, \pi]$ are the estimated posterior probabilities of differential expression and one also has the estimate of ρ so that the size and significance of differences can be assessed.

13.5 A case study: normalization and differential detection

Here we discuss a case study that used microarrays to detect genes that differ in a slightly unusual application in virology first described in Reilly, Wang and Rutherford (2003). Porcine reproductive and respiratory syndrome (PRRS) (European Commission, 1991) is an infectious disease in swine that was initially reported in the United States in 1987 (Keffaber, 1989) and in Europe in 1990. The disease is characterized by late-term abortions, stillbirths, weak piglets, and respiratory problems in pigs of all ages. PRRS causes substantial economic loss in the swine industry. The etiological agent, PRRS virus (PRRSV) was identified shortly after emergence of the disease (Terpstra, Wensvoot and Pol, 1991; Wensvoort et al., 1991). PRRSV predominantly infects and preferentially replicates in pulmonary alveolar macrophages (PAMs) *in vivo* (Rossow et al., 1999; Duan, Nanwynk and Pensaert, 1997; Cheon et al., 1997). PAMs are cells that reside in the lung and are important for proper functioning of the immune system.

In the initial effort to identify genes that impact macrophage function during viral infection, differential display reverse transcription polymerase chain reaction (DDRT-PCR) was used to capture ESTs within PRRSV-infected macrophages over a 24 hour post infection period (Zhang et al., 1999; Wang et al., 2001). From over 8000 DDRT-PCR fragments examined, over 100 porcine-derived DDRT-PCR products induced or inhibited by PRRSV were identified and cloned. The ESTs isolated during DDRT-PCR analysis and certain other known swine genes represent good candidate genes for understanding PRRSV development and spread. Due to technical problems with DDRT-PCR, the results are usually confirmed with another method, such as northern blots or microarrays.

After identification and cloning of these candidate genes, spotted cDNA microarrays were created to explore gene expression in PRRSV infected PAMs at 4 and 24 hours post-infection. The arrays contained 139 ESTs (mostly identified through DDRT-PCR) spotted on each array in triplicate (417 spots per array). The swine PAMs were infected by one of three PRRSV strains of interest: VR2332, Neuro-PRRSV or the Resp-PRRS vaccine strain. In addition, a common reference sample was created by "mock infecting" a set of PAMs with CL2621 cell supernatent, the medium in which the virus is cultured. RNA from the virus infected samples was labeled during re-verse transcription with the red-fluorescent dye Cy5 and mixed with a mock infected sample, labeled with the green-fluorescent dye Cy3, from the same time point. This design economizes on the use of viral infected samples compared to a design that considers all possible 2 way comparisons between the viral infected samples. There were 2 replicates of each of the 6 treatments (2 time points and 3 viral strains), re-sulting in 12 microarrays.

Note that due to the use of DDRT-PCR to screen genes and the small array size, the usual assumption behind most methods for normalization (i.e. not that many genes are differentially expressed) may not be valid. Here we suppose that there are some genes that are unlikely to be differentially expressed across certain array comparisons and we use these genes to normalize the array data. This is possible here because when the arrays were constructed, some genes were spotted on the array that were unlikely to differ in terms of gene expression across the samples.

To account for the need to normalize the array data while attempting to find differ-entially expressed genes, we introduce a simple model and use Bayesian techniques to estimate the parameters in the model. To this end, let x_{1ij} and x_{2ij} denote the log of the intensities for gene i on the jth replicate for dye 1 on arrays 1 and 2, and we let y_{1ij} and y_{2ij} denote the corresponding quantities for dye 2. Then we propose the following model

$$x_{1ij} \sim \text{N}(\phi_1 + \theta_{1i} + \psi_{1i}(\eta_i - \theta_{1i}), \sigma_{1i}^2)$$
$$y_{1ij} \sim \text{N}(\phi_2 + \eta_i, \tau_{1i}^2)$$
$$x_{2ij} \sim \text{N}(\phi_3 + \theta_{2i} + \psi_{2i}(\eta_i - \theta_{2i}), \sigma_{2i}^2)$$
$$y_{2ij} \sim \text{N}(\phi_4 + \eta_i, \tau_{2i}^2)$$

where

$$\psi_{ki} = \begin{cases} 0 & \text{if} & \text{else} \\ 1 & \text{if} & i \text{ is a control gene on array } k. \end{cases}$$

By treating the ψ_{ki} as parameters, we build the possibility that each gene is a control gene for certain scan comparisons into the model. This allows one to specify that certain genes are unlikely to differ across conditions by using a prior that assumes ψ_{ki}

is 1 with some specified probability. For all other parameters (including the ψ_{ki} for genes not known to differ across conditions), one can use standard non-informative priors and use the Gibbs sampler to draw samples from the posterior distribution of all parameters.

Given the probability model and the prior distributions, we can find the posterior distribution of the parameters. This distribution is proportional to

$$\prod_{i=1}^{p}\prod_{j=1}^{2} N(x_{1ij}|\phi_1 + \theta_{1i} + \psi_{1i}(\eta_i - \theta_{1i}), \sigma_{1i}^2)N(y_{1ij}|\phi_2 + \eta_i, \tau_{1i}^2)$$

$$\times N(x_{2ij}|\phi_3 + \theta_{2i} + \psi_{2i}(\eta_i - \theta_{2i}), \sigma_{2i}^2)N(y_{2ij}|\phi_4 + \eta_i, \tau_{2i}^2)/(\sigma_{1i}^2\sigma_{2i}^2).$$

We then obtain credible intervals for the fold changes by drawing simulations from the posterior distribution of all the parameters. The easiest way to draw samples from the probability distribution given above is with the Gibbs sampler. The Gibbs sampler is convenient for this model because all of the conditional distributions we need to sample are either normal ($\phi_k, \theta_{1i}, \theta_{2i}$ and η_i), inverse-gamma (σ_{li}^2 and τ_{li}^2) or Bernoulli (ψ_{li}) (see exercise 5). We then draw from each of these conditional distributions in turn, as explained in section 2.6. One must specify starting values for the parameters that are not simulated first. Here we can just use sample means and variances for the means and variances that appear in the model.

To illustrate the how the method works, while we can use the model for the entire dataset, here we just consider comparing a single Neuro-PRRSV array to a single Resp-PRRS vaccine array. If we use control gene normalization with the gene GAPDH (a gene that is likely constantly expressed as it is involved in basic metabolic tasks that are likely common across most cells) we find many genes differentially expressed (61 up-regulated and 1 down-regulated with at least 90% probability). If we use a non-informative prior for the proportion of genes differentially expressed, we come to very different conclusions. In contrast to the control gene method, we now find 42 genes are up-regulated and 17 down-regulated with probability at least 90%. The reason for the discrepancy is because a central 95% credible interval for the fold change of GAPDH is (0.4, 0.9, 1.7), thus, although this gene is most likely not differentially expressed (as we suspect based on substantive grounds), the best point estimate finds that it is down-regulated, hence if we normalize using this gene all fold changes will be too high. While GAPDH is likely not differentially expressed across conditions, there is uncertainty in its measurement, and the control gene approach is unable to incorporate this uncertainty.

Figure 13.3 displays credible intervals for the log fold changes for all genes spotted on the array. In this figure the lower x-axis shows the log fold changes produced by using the control gene method (so the credible interval for GAPDH is at zero on this axis) while the upper x-axis displays the log fold changes produced using global normalization assuming that the mean logarithm of the level of gene expression is common across arrays (the 2 methods only differ in regard to the placement of zero). The dotted lines indicate cutoffs for a 2-fold change. We see that there are genes

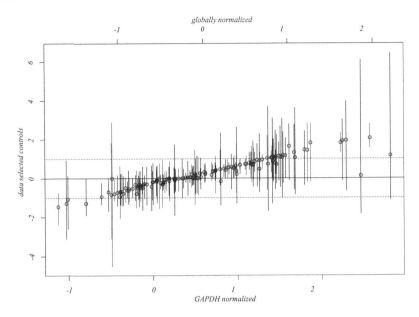

Figure 13.3 *Ninety-five percent probability intervals for the log fold change of a set of genes on a spotted cDNA array obtained from the Bayesian normalization model. The lower and upper axeses give the log fold change using 2 other methods that only differ by an additive constant.*

other than GAPDH that have much narrower probability intervals centered at zero, indicating that there are genes other than GAPDH that act as better controls. These genes act as better controls because they are less variable across arrays compared to GAPDH and do not differ across arrays.

By allowing the data to select the controls, the zero of the scale is placed between the zeros from the control gene method and the global normalization method (despite the fact that neither of the sets of assumptions that justify these methods were employed in the analysis). This indicates that the method that uses global normalization is under-estimating all the fold changes. Figure 13.4 provides insight into the discrepancy between the data selected control method and the global normalization method. In the latter figure, we find histograms of the posterior distribution of the difference in the average intensity (convergence of the chain was also monitored with respect to these quantities). Recall from Section 2.1 that the validity of the global normalization method rests on these histograms being centered at zero, but in fact this is not what is taking place (the probability that the intensity is higher on array 1 exceeds 0.99). Next, if there is no interaction between the dye effect and array, we would expect $(\phi_2 - \phi_1) - (\phi_4 - \phi_3)$ to be centered at zero. In fact, a central 95% credible interval for this difference is (0.67, 1.05), hence there does appear to be an interaction between dye and array. In short, while GAPDH is most likely a control,

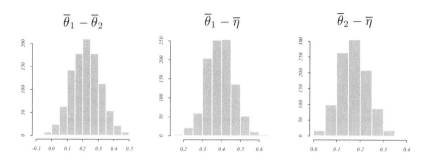

Figure 13.4 *Posterior distributions of differences in average intensity. These histograms should be centered at zero if the assumptions used for global normalization are correct.*

it is not the best control, and the assumptions of the global normalization method are almost surely not satisfied for this comparison.

The application of the method to the entire set of arrays produces interesting results. As an example, consider the swine gene Mx-1 (accession number AF102506). For this gene the 95% credible interval for the fold change in expression for VR2332 relative to Neuro-PRRSV is $(1.4, 4.82)$. This gene is known to be induced by interferon-α (IFN-α) and some viruses. PAMs do not express IFN-α, so viral stimulation is the likely source of expression of this gene. Moreover, the protein produced by this gene confers resistance to influenza in mice. Hence, the fact that this gene is expressed at lower levels in PAMs infected with Neuro-PRRSV may partially account for the ability of this strain to infect tissues VR2332 cannot infect.

13.6 Exercises

1. Show that the MLEs of θ and ϕ_i satisfy $\hat{\theta} = \frac{\sum_i y_i \phi_i}{\sum \phi_i^2}$. Explain how this justifies the last claim in the section on MBEI.

2. How does value of $\frac{1}{2}$ used in SAM relate to how correlated the PM and MM data.

3. Justify the posterior mode claim in the section on the empirical Bayes approach when $a + a_0$ is large.

4. For the empirical Bayes approach, show that if $\theta_g = \theta_r = \theta$ then the marginal likelihood for (a, a_0, ν) is given by the expression

$$p_0(g, r) = \frac{(gr)^{a-1} \nu^{a_0}}{(r + g + \nu)^{2a + a_0}} \frac{\Gamma(2a + a_0)}{\Gamma(a)^2 \Gamma(a_0)}$$

(*Hint:* Replace the 2 independent priors with a single gamma prior since the prior specifies that $\theta_g = \theta_r = \theta$ where θ has a gamma prior).

5. Derive expressions for the conditional distributions of all parameters for the model presented in section 13.5.

6. In a paper by Koinuma et al. (2006), the authors used microarrays to find genes

that are involved in colorectal tumors in subjects with microsattelite instability (MSI). They used Affymetrix microarrays to identify genes that differ among colorectal cancer patients with or without MSI. For each subject, 2 microarrays were used (these are referred to as A and B), and the data is available through the gene expression omnibus (accession number GSE2138). Note that the labels of the arrays indicate the subject identifiers used in the paper, so that one can use the clinical data in Table 1 of that paper to link the clinical data to the microarray data.

(a) Is the array data normalized? Explain.

(b) Use a 2 sample t-test to test for differences between gene expression attributable to MSI status. Do more genes differ than you would expect by chance?

(c) How many genes are found to differ due to MSI status if you control the FDR at 20%?

(d) Repeat the previous part, but now control for differences in age and gender when testing for differences attributable to MSI status. Compare your results to the previous analysis and explain your findings.

(e) Although only spot level summaries are provided, show how one can use SAM to identify genes that differ in expression due to MSI status.

7. In a paper by Wang et al. (2007), the authors use Affymetrix microarrays to find genes in epithelial cells whose level of expression differs due to the level of Lim-only protein 4 (called LMO4). The probe level data is available through the gene expression omnibus (accession number GSE7382).

(a) Using the probe level data, compute MBEI and RMA for all genes on the arrays. Then use these gene level summaries to find genes that differ between groups using a 2-sample t-test (assuming equal variances across the groups) to find genes that differ with an FDR of 10%. Do your results (using either method) agree with the findings the authors present in the paper?

(b) Use SAM to compute probe level summaries and detect differences across genes. Compare to your results from the previous part and to the analysis in the original paper.

8. In a paper by Wade et al. (2008), the authors use Agilent 2 color arrays to test for differences between 2 groups of rats. This data is available in the form of the intensity in each of the channels through the gene expression omnibus (accession number GSE10032).

(a) Are the assumptions that motivated the Bayesian procedure for spotted cDNA arrays in section 13.4 valid for this dataset? If not, normalize the data so that the motivating assumption is valid.

(b) Use the approach from section 13.4 to find genes, if any, whose probability of being differentially expressed is more than 0.95.

(c) Extend the SAM algorithm to detect differences between groups for this experiment and technology. Compare your results to your previous findings.

CHAPTER 14

Cluster Analysis in Genomics

14.1 Introduction

While finding genes that differ between 2 or more classes of samples is frequently the first question addressed in the context of microarray studies, there are other questions of interest. Most of these further issues have to do with examination of the manner in which sets of genes work in concert to perform complex physiological phenomena. For example, multiple genes are typically involved to respond to some extra-cellular signal, hence if a cell receives a signal then all genes involved in the response should have their expression level change. This would induce correlations in the level of gene expression for all genes involved in this cellular response. In addition to finding groups of genes that work together, we may also seek sets of subjects that seem to belong to the same group in terms of their gene expression profiles (i.e. the set of gene expression measurements for all genes). Some methods seek to perform both of these task simultaneously. For concreteness, most of the discussion here will focus on finding groups of genes that behave in similar ways in response to a variety of stimuli.

A number of approaches have been developed to understand the way genes work together, but perhaps the most commonly used of these is a set of techniques called cluster analysis (or unsupervised learning) (see, for example, Eisen, 1998). Given a set of measurements for a collection of units, the goal of cluster analysis is to uncover sets of items that have measurements that are similar. As an example, consider Figure 14.1. This hypothetical example shows the level of gene expression under 2 conditions for 150 genes. The genes represented by triangles have high expression under condition 2 but low expression under condition 1, while genes represented with circles have the opposite property. We expect to find such clusters of genes because their expression is regulated to fulfill some biological function. Hierarchical cluster analysis seeks to determine the relative distances between all items (this information is frequently conveyed via a graphical device called a dendrogram), while nonhierarchical methods seek to partition the items into distinct groups. While we will focus on methods that seek to group items that are similar (i.e. minimize the within cluster distance between items), there are also approaches based on the idea of making distinct clusters highly dissimilar, or a combination of high dissimilarity between

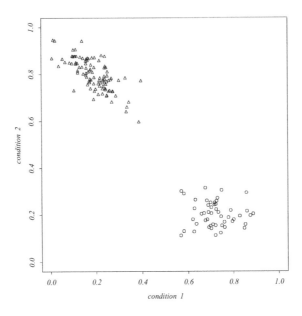

Figure 14.1 *An example of data that exhibits clusters. Each point corresponds to the expression level of a gene and the 2 axeses represent 2 different conditions.*

clusters and high similarity within clusters. There are a number of good textbooks devoted to cluster analysis, such as Hartigan (1975) and Gan, Ma, and Wu (2007).

If we seek to group similar items, then we must decide what similar means. Many methods in popular use require the definition of a dissimilarity measure, either between items, between an item and a cluster or between clusters. If we can define these dissimilarity measures in a relevant way, then we are faced with the problem of determining the allocation of items to clusters in a manner that minimizes the within cluster distance between items. For a fixed number of clusters, this is just a combinatorial problem: what is the optimal way to assign units to a fixed number of clusters. Unfortunately, in most applications the number of ways of allocating items to clusters is too large to consider every possible allocation, hence even if one has a satisfactory measure of item and cluster dissimilarity there is the computational problem of determining the optimal allocation of items. While these 2 problems (i.e. item dissimilarity measures and finding the optimal allocation) may appear to be separate, there is considerable interaction between these 2 components of the problem.

Most of the methods that have been proposed to examine coregulation of gene expression that don't depend directly on a clustering algorithm use the correlation among genes across replicates to find sets of genes that have large correlations among them. Some have used the singular value decomposition to find sets of genes that contribute substantially to the observed correlation structure, such as Alter, Brown and Botstein (2000), while others have used the closely related principal components decomposition, such as Yeung and Ruzzo (2001). Relevance networks (Butte and Kohane, 2000) apply simple thresholds to the observed absolute value of pairwise correlations to identify sets of coregulated genes (although the authors suggest that other measures of pairwise similarity are also useful).

14.1.1 Dissimilarity measures

There are many similarity and dissimilarity measures in common use. If we let x_i represent the collection of gene expression levels for gene i across all arrays (so if there are n arrays, x_i is an n-vector), then the Minkowski distance given some integer m, is given by the expression

$$d(x_i, x_{i'}) = \left(\sum_{j=1}^{n} |x_{ij} - x_{i'j}|^m \right)^{\frac{1}{m}}.$$

Frequently m is set to 2 so that the Minkowski distance just becomes the usual Euclidean distance. Another common distance called the Manhattan distance is obtained by setting $m = 1$. Correlation is the most commonly used measure of item similarity.

14.1.2 Data standardization

Another consideration in specification of dissimilarity measures concerns standardization of the data set. For example, in the context of microarray studies, if a pair of genes were measured over the life cycle of a cell and always moved in the same direction as the cell cycle progressed, then one would be inclined to think that such genes are involved in the same pathway, and/or regulated by the same set of other genes. One would like a clustering method to classify such genes as in the same cluster, however, if one of these genes is always expressed at a relatively high level whereas the other gene is always expressed at a low level, then the Euclidean distance between the pair can be rather high. Indeed such genes are likely to found to be more dissimilar than a pair of genes that are almost always constantly expressed at some moderate level. A simple remedy to this is to subtract the mean level of gene expression over all conditions for each gene from the level of expression from each condition so that each gene has zero mean expression over all conditions. Similarly, it frequently makes sense to standardize all the gene expression values so that the standard deviation across all conditions is 1. This guards against the scenario where 2 genes that stay nearly constant across all conditions are found to be closer than 2 genes that take large swings over conditions but move together across all conditions.

14.1.3 Filtering genes

Many of the algorithms that have been introduced do not work well in situations where there are many items that are to be clustered, that is, when there are many genes. The same is true for classification algorithms, which are discussed in Chapter 15. In the context of microarray analysis there are frequently tens of thousands of genes under consideration, and this is sufficiently large for many algorithms to work very slowly and or have difficulty finding the most reasonable solution. For this reason, many researchers first use a method for *filtering* the genes. This filtering consists of finding sets of genes that are likely to be of not much use in defining clusters of genes. For example, in the context of using Affymetrix arrays, we may seek to only find clusters of genes that are expressed in the sample, so we may use the presence call to filter the genes. For example, we may require that a gene be found present on every array. In other contexts, we may only want to use genes whose standard deviation across replicates is sufficiently small or whose average level of gene expression is above some threshold. The idea behind all these approaches is that we shouldn't just be finding clusters driven by random error.

14.2 Some approaches to cluster analysis

Many methods have been proposed for cluster analysis. Here we discuss some of the most popular methods in the microarray literature, namely hierarchical, center-based and model based. A number of other methods have been proposed in the rather extensive cluster analysis literature, and a discussion of some of these can be found in a number of texts, such as Gan, Ma and Wu (2007).

14.2.1 Hierarchical cluster analysis

Hierarchical clustering methods either successively divide up the units (divisive hierarchical clustering) or successively merge units (agglomerative hierarchical clustering) until no further divisions or mergers of units are possible. Unlike other methods for clustering items, there is no need to select a number of clusters. Instead, the result from the algorithm is just a set of merges or splits of all items. This set of merges or splits is usually presented graphically via a dendrogram. These methods require the user to define dissimilarity measures between individual items, and from these dissimilarity measures there are a number of standard ways for defining dissimilarity measures between pairs of clusters (which is required for the algorithm).

In addition to a dissimilarity measure for units, to conduct hierarchical cluster analysis one also must specify dissimilarity measures for clusters. There are 3 common methods to define distances between clusters in the context of hierarchical cluster analysis, and these are referred to as linkage methods: single linkage, average linkage and complete linkage. Given 2 clusters of objects, if we use $d(x_i, x_j)$ to represent the dissimilarity measure between items x_i and x_j and $\delta(U, V)$ to represent the dissimilarity measure between cluster U and V, then these cluster dissimilarity measures can be expressed

1. single linkage $\delta(U, V) = \min_{x_i \in U, x_j \in V} d(x_i, x_j)$

2. average linkage $\delta(U, V) = \frac{1}{\sum_{x_i \in U, x_j \in V}} \sum_{x_i \in U, x_j \in V} d(x_i, x_j)$

3. complete linkage $\delta(U, V) = \max_{x_i \in U, x_j \in V} d(x_i, x_j)$.

Since single linkage uses the shortest distance between 2 clusters it has a tendency to produce long thin clusters compared to the other 2 approaches, hence it is unusual to use single linkage in a microarray study. Note that for a given number of clusters, in principle one could determine the dissimilarity measure between all clusters, and so in principle, one could find the best allocation of items to clusters. In practice there are usually far too many ways to allocate items to clusters, hence a number of techniques have been proposed that are suboptimal, but are fast and frequently of practical use.

The basic workings of the algorithm used for agglomerative hierarchical clustering are quite simple. If one has p items to cluster, one starts with p clusters, and the distance between clusters is just the dissimilarity between the individual items. One then finds the 2 clusters that are closest and merges these 2 clusters so that there are only $p-1$ clusters. Then one recomputes the distance between this new cluster and all other clusters (the dissimilarity between all other clusters that were not merged stays the same). Then one repeats the first step: find the pair of clusters that are closest and merge these to create a new cluster. This continues until there is only one cluster left. Note that if there is the same value for the dissimilarity measure between 2 items it is possible that there is a tie as the algorithm proceeds, and this leads to some ambiguity regarding the order of the mergers. Hence there is not always a unique set of mergers for a given data set, but the graphical summary of the hierarchical clustering process, the dendrogram, will still be unique.

Divisive clustering is conceptually similar except that one starts with a single cluster, then breaks this into 2 clusters so that the clusters are maximally separated. Unfortunately for even a moderately sized cluster, there are too many ways to split the cluster into 2 clusters, hence divisive algorithms must consider only a subset of all possible splits. Further steps split up a cluster to create 2 more clusters. This creates even more problems as there are many clusters to split and introduces the problem of monotonicity. This is the idea that the algorithm should not split up a subcluster of some parent cluster prior to splitting up the parent cluster further. Nonetheless, algorithms have been designed that are monotone and practical algorithms have been devised (Kaufman and Rousseeuw, 1990).

A dendrogram is a tree-like structure (actually it is upside down tree-without roots) used to graphically represent the series of mergers that define the hierarchical clustering result. In Figure 14.2 a dendrogram is shown that results from using complete linkage on the data from Figure 14.1. At the base of the figure is a dot for each item and on the left margin is an axis giving the level of the dissimilarity measure which we tolerate in allowing items to belong in the same cluster. Hence at the level of 0 we don't allow any dissimilarity between objects in the same cluster, so all items are in their own cluster. As we increase the tolerated level of dissimilarity for objects in the same cluster, clusters merge and there are fewer groupings of objects. For the

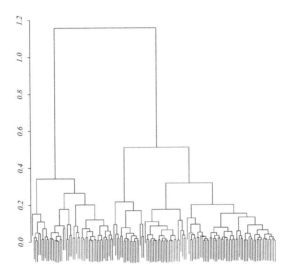

Figure 14.2 *A dendrogram for the data from Figure 14.1.*

example in Figure 14.1 we find that the algorithm prefectly separates the 2 groups when it makes the first split (although this is an illustrative example and not typical of actual data sets).

All one needs to perform a cluster analysis is a set of similarities of dissimilarities, hence here, as an example, we consider clustering amino acids on the basis of the BLOSUM50 substitution matrix using complete linkage. We can think of each amino acid as residing in a 20 dimensional space where similar coordinates in this space indicate how similar a set of amino acids are in terms of their ability to substitute for each other and other amino acids. We use Euclidean distance in this space, hence at the start of the algorithm there are 20 clusters, where the distance between the clusters is determined by the Euclidean distance between the vectors that are the rows of the BLOSUM50 substitution matrix given in Table 7.1. We then combine the 2 closest of these clusters (namely I and V, both nonpolar neutral residues) and compute the distance between this new cluster and all remaining clusters (all the other cluster distances remain constant). Continuing in this fashion, one obtains the dendrogram presented in Figure 14.3. Note that the pairings obey the amino acid classifications given in the first chapter until the level of dissimilarity between Q and E is reached, but these 2 residues have very similar chemical structures. Similarly, the pairs N and D and F and Y, while in different classes of amino acids, also have very similar chemical structures. Furthermore, the residues that depart from the others the greatest, namely C and W have properties that make them quite different from the others (C forms disulfide bonds while W has one more ring of carbons than any other residue).

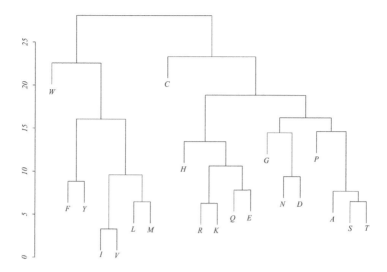

Figure 14.3 *An example of hierarchical clustering. Here the amino acids are clustered using the BLOSUM50 substitution matrix. Letters indicate the single letter designators of the amino acids.*

14.2.2 K-means cluster analysis and variants

A very popular and fast algorithm for clustering items is called k-means, first proposed by MacQueen (1967). Unlike hierarchical clustering, one needs to specify the number of clusters to use this algorithm. If there are k clusters, then one starts the algorithm by determining k centers for the clusters. Then each item is assigned to the cluster with the closest cluster center in terms of the distance measure used (in the usual implementation of k-means, Euclidean distance is used). One then computes the centroid of each resulting cluster (by finding the mean of the coordinates of all items in that cluster). These centroids are then the new cluster centers, and one again assigns items to the clusters based on distance to the cluster center. When no items are assigned to new clusters the algorithm stops. One could use this to produce hierarchical clusters by then clustering the cluster centers, and by using a decreasing sequence for the number of clusters one could produce a hierarchical clustering solution.

The set of initial cluster centers has a large impact on the results, hence one should use many sets of cluster centers in practice and use the clustering solution that is optimal in some specific way. For example, we may want to minimize the average within cluster distance between items, hence at the end of each run of the k-means algorithm, we would compute this quantity. Another problem that can arise is that some cluster may have no objects in it at some point and so the algorithm is poorly defined. This usually only occurs if one is using many clusters relative to the number of items being clustered or if one selects cluster centers far from the locations of all the items. This can usually be avoided by using the locations of items as the initial cluster centers. A simple approach is to use a large number of random samples from the set of all items of size k, and use each of these as initial values. Then one can compare the quality of the solutions using some criterion, such as the average within cluster sum of squares. There are a number of extensions and variations on this algorithm. For example, k-medoids uses the point whose average dissimilarity to all items as measured by some median distance in the cluster is minimal. K-medoids is more robust to outliers than k-means. Partitioning around medoids (PAM) is a popular implementation of this idea that is widely used in the analysis of microarrays. Note that the difficulty here is that there will not generally be any observation that is the median with respect to all of the variables that one is using to find clusters.

14.2.3 Model based clustering

In addition to the algorithms just discussed, there is a way to treat the clustering problem in a model based fashion. This approach was first developed by Banfield and Raftery (1993) and applied to the analysis of microarrays in Yeung et al., (2001). The advantages of this approach are that the focus now becomes the adequacy of the model and this can be assessed with model diagnostics or model selection criteria. With the previous approaches this isn't really feasible since these other approaches are just algorithms, and this leads to selection of the clustering algorithm via *ad hoc* choices or basing the results on some test set (which may or may not be relevant for a new data set that one is analyzing). Another advantage is that one can treat the number of clusters as a parameter and estimate that parameter from the data (at least under certain conditions). Finally, use of a rigorous statistical model allows one to include uncertainty estimates of cluster identity and descriptions of the clusters that are not feasible in the context of approaches based on a given algorithm.

The fundamental insight that is the basis for model based clustering is that finding clusters for observations which are n-vectors is equivalent to fitting mixture models for observations in an n dimensional Euclidean space. For example, suppose $n = 1$ so that for each gene we have just a single measurement, and further suppose that p_1 genes have expression levels that are around -2 (say between -1 and -3) and p_2 gees have expression levels that are around 2 (say between 1 and 3). In this simple example, one is inclined to say that there are 2 clusters (one centered near -2 and one centered near 2). To make this more precise, we could suppose that the location of each measurement is distributed according to a 2 component normal mixture model: $x_i \sim \lambda N(\mu_1, \sigma^2) + (1 - \lambda)N(\mu_2, \sigma^2)$ (choosing a common variance for simplicity here). For the simple example, reasonable estimates of the parameters would just be $\hat{\mu}_1 = -2$, $\hat{\mu}_2 = 2$, $\hat{\sigma} = .5$, and $\hat{\lambda} = \frac{p_1}{p_1+p_2}$. In practice, one can estimate the parameters using maximum likelihood or posterior modes. Similarly, the data displayed in Figure 14.1 could be interpreted as 150 draws from a 2 component bivariate normal mixture distribution with one component centered at $(0.2, 0.8)$ and the other centered at $(0.8, 0.2)$.

More formally, suppose there are p units we seek to cluster and we have n observations on each unit. Let $\gamma_i = k$ is unit i is from the kth cluster. If we use θ to represent the set of parameters for all mixture components (in the previous 2 component mixture model example, $\theta = (\mu_1, \mu_2, \sigma)$) and if the kth mixture component has density $f_k(x|\theta)$, then we can write the likelihood for the data $x_1, \ldots, x_p$ as

$$L(x_1, \ldots, x_p|\theta, \gamma) = \prod_{i=1}^{p} f_{\gamma_i}(x_i|\theta).$$

To make this expression more useful we will suppose $f_k(x|\theta)$ is an n-dimensional multivariate normal distribution with mean μ_k and covariance matrix Σ_k. This implies $f_k(x|\theta) = (2\pi)^{-\frac{n}{2}} |\Sigma_k|^{-\frac{1}{2}} \exp\{-\frac{1}{2}(x - \mu_k)'\Sigma_k^{-1}(x - \mu_k)\}$. For now, we will assume there are a total of K clusters. If we introduce the notation E_k to represent the set of i so that $\gamma_i = k$, then we can write the likelihood as

$$L(x_1, \ldots, x_p|\theta, \gamma) \propto \prod_{k=1}^{K} \prod_{i \in E_k} |\Sigma_k|^{-\frac{1}{2}} \exp\left\{-\frac{1}{2}(x_i - \mu_k)'\Sigma_k^{-1}(x_i - \mu_k)\right\},$$

where we have ignored the constant involving π that appears in the expression for the multivariate normal distribution.

Next, note that if one knows the vector γ, since the MLE of the mean of a normal distribution is the sample mean, if $p_k = \sum_i 1_{\{\gamma_i = k\}}$, then the MLE of μ_k is $\overline{x}_k = p_k^{-1} \sum_{i \in E_k} x_i$. If we then substitute this expression for the MLE of μ_k into the expression for the likelihood and take the logarithm, we find that the log-likelihood can be written for some constant C

$$\ell(x_1, \ldots, x_p|\theta, \gamma) = C + \sum_{k=1}^{K} \sum_{i \in E_k} \left[-\frac{1}{2} \log |\Sigma_k| - \frac{1}{2}(x_i - \overline{x}_k)'\Sigma_k^{-1}(x_i - \overline{x}_k) \right].$$

If trace(A) is used to denote the trace of the matrix A (i.e. the sum of the elements on the diagonal of A), then if y and z are 2 n-vectors and A is an n by n matrix it is not hard to show that $y'Az = \text{trace}(zy'A)$ so that we can write the log-likelihood as

$$\begin{aligned}
\ell(x_1, \ldots, x_p|\theta, \gamma) &= C + \sum_{k=1}^{K} \sum_{i \in E_k} \left[-\frac{1}{2} \log |\Sigma_k| \right. \\
&\quad \left. - \frac{1}{2}\text{trace}\left((x_i - \overline{x}_k)(x_i - \overline{x}_k)'\Sigma_k^{-1}\right) \right] \\
&= C - \frac{1}{2} \sum_{k=1}^{K} \left[p_k \log |\Sigma_k| \right. \\
&\quad \left. + \sum_{i \in E_k} \text{trace}(x_i - \overline{x}_k)(x_i - \overline{x}_k)'\Sigma_k^{-1}) \right].
\end{aligned}$$

If we then define $W_k = \sum_{i \in E_k} (x_i - \overline{x}_k)(x_i - \overline{x}_k)'$ we find that the log-likelihood can be expressed as

$$\ell(x_1,\ldots,x_p|\theta,\gamma) = C - \frac{1}{2}\sum_{k=1}^{K}\left[p_k\log|\Sigma_k| + \text{trace}(W_k\Sigma_k^{-1})\right].$$

While one can attempt to maximize this expression with respect to the vector γ and Σ_k for $k = 1,\ldots,K$, frequently further assumptions are introduced to simplify the computational and inferential burden by reducing the number of parameters. Assumptions about Σ_k for $k = 1,\ldots,K$ are equivalent to assumptions regarding the sorts of clusters we expect to find. To understand the relationship, think about what samples from the multivariate normal distribution implied by the set of Σ_k's would look like. For example, if we set $\Sigma_k = \sigma^2 I$ for all k (where I is an n by n identity matrix), then sample realizations from the normal likelihood would appear as observations from a normal distribution with uncorrelated components. In Figure 14.1 for example, it appears that this data could have arisen as a set of iid observations from a 2 component mixture model where there is a negative correlation in the cluster represented by triangles and there is no correlation in the other cluster. Hence, if we seek spherical clusters then we find upon substituting $\Sigma_k = \sigma^2 I$ into the previous expression for the log-likelihood that

$$\begin{aligned} \ell(x_1,\ldots,x_p|\theta,\gamma) &= C - \frac{1}{2}\sum_{k=1}^{K}\left[p_k\log(\sigma^2) + \sigma^{-2}\text{trace}(W_k)\right] \\ &= C - \frac{1}{2}\left[p\log(\sigma^2) + \sigma^{-2}\sum_{k=1}^{K}\text{trace}(W_k)\right] \end{aligned}$$

so that we find the MLE of γ by selecting $\hat{\gamma}$ to minimize just the last term, which implies we must minimize $\sum_{k=1}^{K}\text{trace}(W_k) = \text{trace}(W)$ where $W = \sum_{k=1}^{K}W_k$. The matrix W plays a key role in a number of clustering methods because it measures the within cluster sum of squared deviations from the cluster center averaging over all clusters. If we assume that the clusters are spherical, then we ignore all elements of W which are on the off diagonal which is sensible since we have assumed all these values are zero. This is the basis for a clustering algorithm called Ward's hierarchical agglomerative procedure (Ward, 1963), an algorithm proposed independently of the model based approach to cluster analysis presented here.

If we instead assumed that $\Sigma_k = \Sigma$ for $k = 1,\ldots,K$ where we only assume Σ is positive definite, then the clusters would no longer have spherical shapes, although we still assume all clusters have the same shape. Under this assumption, the log-likelihood becomes

$$\begin{aligned} \ell(x_1,\ldots,x_p|\theta,\gamma) &= C - \frac{1}{2}\sum_{k=1}^{K}\left[p_k\log|\Sigma| + \text{trace}(W_k\Sigma^{-1})\right] \\ &= C - \frac{1}{2}\left[p\log|\Sigma| + \sum_{k=1}^{K}\text{trace}(W_k\Sigma^{-1})\right] \end{aligned}$$

but the sum of the traces of a set of matrices is just the trace of the sum of those matrices, so that

$$\ell(x_1,\ldots,x_p|\theta,\gamma) \;=\; C - \frac{1}{2}\left[p\log|\Sigma| + \text{trace}(W\Sigma^{-1})\right].$$

If we then maximize this expression with repect to Σ we find that the MLE of Σ is just W/p (see exercise 4). Note that this MLE is a function of γ (which is unknown), but if we then substitute this expression into the previous expression for the log-likelihood, we find that

$$\begin{aligned}
\ell(x_1,\ldots,x_p|\theta,\gamma) &=\; C - \frac{1}{2}\left[p\log|W/p| + \text{trace}\left(W\,(W/p)^{-1}\right)\right] \\
&=\; C - \frac{1}{2}\left[p\log|W| - p\log(p) + np\right]
\end{aligned}$$

so that we maximize the likelihood by finding the value of γ so as to minimize $|W|$. Once again, our method uses W to find the best assignment of units to clusters, although now we use the determinant of this matrix. This method has also been proposed prior to the model based approach to cluster analysis (this is the criterion of Friedman and Rubin, 1967).

While one can in principal make no assumptions regarding the form for Σ_k for $k = 1,\ldots,K$, in practice one frequently obtains clusters that are difficult to interpret and makes computation more difficult. This is because one must use a numerical procedure to find local modes of the likelihood. In practice, the EM algorithm can be used to find local modes for mixture models (a simple example with only 2 components was presented in Section 13.4). In addition, frequently the data do not support a single best MLE for γ because the clusters overlap in n-dimensional space. In this case there can be a lot of uncertainty with regard to which cluster a given unit belongs. This situation leads to the existence of many local modes for the likelihood. The existence of many such local modes makes finding the global mode very difficult, thus leading any method that seeks a local mode to converge to a location in parameter space that is not the global mode. One can also treat the number of clusters as a parameter and attempt to determine its value based on the dataset. The authors originally proposed a method based on Bayes factors, but this doesn't always give satisfactory results.

14.3 Determining the number of clusters

Determining the number of clusters in a data set is a difficult and widely studied problem. Some of the approaches to clustering that we have discussed require the specification of the number of clusters, for example, k-means. Many approaches have been proposed, so here we will simply outline the primary ideas that have been proposed. Milligan and Cooper (1985) conducted a large scale simulation study in which many methods for determining the number of clusters were considered. We will discuss some of the methods that were found to perform the best in that study, in addition to a method that was proposed more recently.

It is not clear how many clusters to expect in a genomics study. We typically think of the number of clusters as corresponding to the number of distinct pathways operating in the system under investigation, however the level of specificity of a pathway is difficult to discern when one discusses such pathways. For example, in the context of the GO, genes are modeled as being in a functional hierarchy, so that if one thinks of very detailed pathways, one could have scores or hundreds of clusters, whereas at a less refined resolution there would perhaps be as few as a dozen clusters. Some biologists tend to think of the number of clusters as determined by the number of distinct conditions that one considers. For example, one may think there is a group of genes that are upregulated in response to some stimulus (i.e. the level of gene expression increases), another group that are downregulated and the rest remain unchanged. Such a situation would give rise to 3 clusters. As the number of conditions under which one has data increases, the number of potential clusters would thereby increase exponentially. Here we say potential clusters since it is possible that there is overlap in the sets of genes that are altered by different conditions.

To understand the nature of the difficulty of determining the appropriate number of clusters, suppose one is using k-means and one wants to minimize the average distance between items in a cluster. If we treat the number of clusters as a parameter and choose this parameter to minimize the average within cluster distance, then the optimal number of clusters is equal to the number of items one is trying to sort into clusters, and the locations of the cluster centers are given by the locations of the items one is clustering. Such a clustering solution would yield zero average within cluster distance. Similar considerations apply to hierarchical and model based clustering. Moreover, constraints on the size of a cluster, for example, that each cluster contains at least m items, are of no assistance in the sense that the optimal number of clusters will be determined by the constraint rather than the data. While this fact is troubling, certainly some method should be able to recognize that there are 2 clusters in Figure 14.1, and indeed, this is the case.

Central to a number of methods is the average within cluster distance between items in the same cluster. If we let D_k represent the total distance between items in cluster k,

$$D_k = \sum_{i,j \in E_k} d(x_i, x_j)$$

and set

$$W_K = \sum_{k=1}^{K} \frac{1}{2p_k} D_k$$

then W_K measures the extent to which clusters are similar when there are K clusters. If d is the squared Euclidean distance, then W_K is average within cluster sum of squared centered around the cluster means (see exercise 5). While W_K monotonically decreases as K increases, frequently in the presence of clustered data, W_K

will drop sharply as K increases for some value of K. In such cases a reasonable approach is to choose the value of K so that $|W_K - W_{K-1}|$ is large.

Several methods use W_K in a more formal fashion to determine the number of clusters. For instance Krzanowski and Lai (1985) assume that d is the squared Euclidean distance and define

$$V(K) = (K-1)^{2/n} W_{K-1} - K^{2/n} W_K$$

(recall n represents the number of measurements on each unit, i.e. the number of arrays). One then chooses the number of clusters to maximize

$$C_1(K) = \left| \frac{V(K)}{V(K+1)} \right|.$$

More recently, Tibshirani, Walther and Hastie (2001) used W_K in a slightly different fashion. They define

$$G(K) = \mathrm{E}\left[\log(W_K)\right] - \log(W_K)$$

and choose K to maximize this expression. This statistic is called the gap statistic. Here, $\mathrm{E}\left[\log(W_K)\right]$ represents the expectation of $\log(W_K)$ under a sample of size p from a population where $K = 1$. One can most simply compute this expectation using a simulation technique in which one simulates a data set of size p by drawing each sample independently of the others, and simulating each sample by drawing each measurement from a uniform distribution over the range of values observed for each measurement. Suppose that L such simulations are conducted. One then replaces the expectation in the definition of G with the sample average of the values of $\log(W_K)$ from these simulations. To account for the variability in this estimate, one also computes the standard deviation of $\log(W_K)$, denoted s_K, sets $\tilde{s}_K = s_K \sqrt{1 + 1/L}$ then selects K so that K is the smallest number so that

$$G(K) \geq G(K+1) - \tilde{s}_{K+1}.$$

Hence this approach is based on finding large changes in W_K as K varies, however this method accounts for the sampling variability in W_K. Moreover, the authors demonstrate that the simulation technique is optimal in that it is least likely to give support for the existence of more than 1 cluster when in fact there is only 1 cluster.

A similar approach has been proposed by Calinski and Harabasz (1974), except they consider the distance between clusters in addition to the within cluster distance. If we let $W(K)$ represent the within cluster sum of squares and $B(K)$ the between cluster sum of squares (defined by the distance of the cluster centers to the overall center of the data), then Calinski and Harabasz choose K to maximize

$$C_2(K) = \frac{B(K)/(K-1)}{W(K)/(p-K)}.$$

Note that both $C_1(K)$ and $C_2(K)$ are undefined for $K = 1$ so that they always find support for $K > 1$. In contrast, $G(K)$ is defined for all $K \geq 1$ so that the gap statistic can be used to test for the presence of any clusters in a data set.

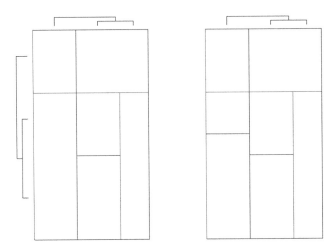

Figure 14.4 *An example of partitioning a data matrix with respect to the rows and columns. Note that for the partition on the left, there is a dendrogram that corresponds to the partition for both the rows and the columns, but for the partition on the right there is no dendrogram for the rows.*

14.4 Biclustering

While our discussion has thus far focused on finding clusters of genes that display the same behavior across different conditions (where these conditions are arrays), we could employ the same set of methods to find clusters of subjects that display the same behavior across a large set of genes. In fact, one can even attempt to cluster subjects and genes simultaneously, a process called biclustering. While a number of algorithms have been proposed (see, for example, Kluger et al., 2003), we will consider the method introduced by Hartigan (1972) here as it demonstrates some of the complexities in biclustering a data set.

In Hartigan's approach, a cluster is a submatrix of the $p \times n$ data matrix. Note that given a set of submatrices of a data matrix, we can cluster both the columns and the rows separately by putting an item in the cluster that it maximally intersects (ignoring ties). However, we cannot uniquely define a dendrogram for the rows and the columns based on an arbitrary collection of submatrices. To see this, consider Figure 14.4 which shows 2 partitions of a data matrix. For the partition on the left there is a dendrogram for the rows and columns that corresponds to the partition, however for the partition on the right there is no corresponding dendrogram. Hartigan's approach restricts the biclustering solution such that there are dendrograms to describe the clustering of the items on the rows and columns. Not all approaches use this restriction.

Since we are clustering variables of a similar nature (i.e. gene expression for similar biological samples), it is natural to seek submatrices for which the response is constant on average. Hence the goal is to find a partition of the data matrix so that responses are approximately constant within an element of the partition. The algorithm proceeds by entering splits which separate one of the blocks in the current partition (the data matrix is the initial partition). Let A_{ij} represent the value at position i, j in the data matrix and let $B_1, \ldots, B_k$ be the set of blocks at the kth stage of the algorithm. Each block, B_l, is a set of contiguous rows R_l and columns C_l (the number of rows is r_l and the number of columns is c_l). At each stage of the algorithm, B_l is split into 2 other blocks at either a row, to give $B_l^1 = (R_l^1, C_l)$ and $B_l^2 = (R_l^2, C_l)$ (with rows r_l^1 and r_l^2 respectively), or a column, to give $B_l^1 = (R_l, C_l^1)$ and $B_l^2 = (R_l, C_l^2)$. To measure the quality of a partition, we introduce the notation that $A(B_l)$ is the mean of all elements in block B_l and we use the sum of the squared errors in each block

$$SS = \sum_l \sum_{i,j \in B_l} \left(A_{ij} - A(B_l) \right)^2$$

to measure the quantity of our clustering solution. The change in the sum of squares due to splitting B_l at some row to get B_l^1 and B_l^2 will be

$$\Delta SS = c_l r_l^1 \left(A(B_l^1) - A(B_l) \right)^2 + c_l r_l^2 \left(A(B_l^2) - A(B_l) \right)^2.$$

If some row split has already occured, then the row splits are restricted since we want to have a dendrogram for the rows and columns (such a split is called a "fixed split"). When we aren't faced with this constraint (which is called a "free split"), we should multiply the ΔSS by $\pi/(2r_l)$ since we can obtain a larger reduction just due to chance than if we are constrained by previous splits. The value $\pi/(2r_l)$ is derived by comparing the change in ΔSS we expect when there are no differences for free splits.

Splits continue until ΔSS is what we expect based on chance. Again, there are complications introduced due to the free and fixed splits. If we let SS_1 represent the reduction in the sum of squares due to N_1 free splits, let SS_2 represent the reduction in the sum of squares due to N_2 fixed splits, and let SS_3 be the sum of squared deviations within clusters supposing that N_3 is the total number of responses (i.e. the number of elements in A which is np) less the number of clusters, then the algorithm stops when

$$\frac{SS_3}{N_3} > \frac{SS_1/2 + SS_2}{N_1/\pi + N_2}.$$

This criterion follows from considering the distribution of ΔSS under the null of the existence of no clusters. We further assume that elements of the matrix A are iid normal. If $\overline{x}_1$ is the mean of all responses A_{ij} for $i \leq k$ and $\overline{x}_2$ is the mean of all responses A_{ij} for $i > k$, for fixed splits $\Delta SS = (\overline{x}_1 - \overline{x}_2)^2 nk(p-k)/p$ and this is distributed as a $\sigma^2 \chi_1^2$ random variable for some variance parameter σ^2. For the free splits, we are maximizing over all possible splits, and it transpires that we have the approximation that ΔSS is distributed as a $2\sigma^2 \chi_{p/\pi}^2$ random variable. This implies that $SS_1 \sim 2\sigma^2 \chi_{N_1/\pi}^2$ and $SS_2 \sim \sigma^2 \chi_{N_2}^2$. Finally we have $SS_3 \sim \sigma^2 \chi_{N_3}^2$ since we are just computing the variance of a set of iid random normal deviates. Using the fact that the expected value of a χ^2_m random variable is just m we then find that

$$E[SS_1/2 + SS_2] = \frac{\sigma^2 N_1}{\pi} + \sigma^2 N_2$$

or that

$$\frac{E[SS_1/2 + SS_2]}{N_1/\pi + N_2} = \sigma^2.$$

But then noting that

$$ESS_3/N_3 = \sigma^2$$

we find that we can compare these last 2 expressions and we find that the stopping rule stops finding new clusters when further splits will not on average reduce the error.

One problem with this algorithm is that it will most likely not put 2 rows in the same block if they are separated by many other rows that belong in a different cluster. This is unfortunate as the order of the rows and columns should not impact the clustering solution. In practice the effect of this can be mitigated by first applying a hierarchical clustering algorithm to the rows and columns separately, then using these cluster solutions to help order the rows and columns prior to running the biclustering algorithm.

14.5 Exercises

1. Is correlation a linear function of Euclidean distance? Explain.

2. In the context of the model based approach to cluster analysis, suppose we wanted to allow the orientations of the clusters to vary while keeping the size and shape constant. We can enforce this by letting

$$\Sigma_k = \Phi_k \Lambda_k \Phi'_k$$

where Φ_k is a matrix of the eigenvectors of Σ_k and Λ_k is a diagonal matrix holding the eigenvalues of Σ_k (so this is what is known as the spectral decomposition of Σ_k), and requiring $\Lambda_k = \Lambda$ for all k (here Λ is specified to give the desired size and shape). If we let the spectral decomposition of W_k be written

$$W_k = L_k E_k L'_k$$

then show that the optimal clustering solution is found by minimizing

$$S = \sum_k \text{trace}(\Lambda^{-1} E_k).$$

3. Show that if y and z are 2 n-vectors and A is a n by n matrix then $y'Az = \text{trace}(zy'A)$.

4. Show that when one has p observations from an n dimensional multivariate normal distribution where $p > n$ the MLE of the covariance matrix is given by the sample covariance matrix (*Hint:* Use the fact that the determinant of a matrix is

the product of its eigenvalues and the trace is the sum of its eigenvalues, and find the MLE of $S\Sigma^{-1}$ where $S = \frac{1}{n}\sum_i(x_i - \bar{x})'(x_i - \bar{x}))$.

5. Show that W_k is the average within cluster sum of squared centered around the cluster means when d is the squared Euclidean distance.

6. In the context of Hartigan's approach to biclustering, show that for fixed splits $\Delta SS = (\bar{x}_1 - \bar{x}_2)^2 nk(p - k)/p$ and this is distributed as a $\sigma^2\chi_1^2$ random variable for some variance parameter σ^2.

7. In Gasch et al. (2000), the authors investigate stress responses in *S. cerevisiae*. In one experiment, they used spotted cDNA arrays to measure responses to amino acid starvation over time. In another experiment, they examined the response to a rapid increase in temperature over time. You can find both of these datasets by searching in the gene expression omnibus. The accession number for the submission is GSE18 of both experiments, and the individual data sets are listed as the GDS 115 record for the amino acid and adenine starvation time course experiment, and the GDS 112 record for the heat shock from 30° C to 37° C time course experiment.

(a) Produce a dendrogram using agglomerative hierarchical clustering and complete linkage for both data sets to find clusters of genes without standardizing the data in any manner (use the normalized log ratios). Repeat this exercise but first standardize the data so that each gene has zero expression over time and standard deviation 1. Compare the results of the 2 approaches.

(b) Use k-means clustering to find clusters of genes for both of these datasets using 20 clusters. Are there similar groups of genes identified in both experiments?

(c) Use the method of Krzanowski and Lai to determine the number of clusters one should use in the amino acid starvation experiment if one uses k-means.

(d) Use the gap statistic to determine the number of clusters one should use in the amino acid starvation experiment if one uses k-means. Does the result agree with the method of Krzanowski and Lai?

(e) Use model based clustering with spherical clusters to find clusters of genes using both datasets simultaneously and select the number of clusters using the gap statistic. Discuss the role of data standardization and how it impacts your findings.

CHAPTER 15

Classification in Genomics

15.1 Introduction

The term classification (also known as supervised learning) refers to a set of techniques that are designed to construct rules that allow units to be sorted into different groups. These rules are based on data and the rules are designed to optimize some fixed criterion. In the context of microarray analysis, classification could potentially be useful for medical diagnosis. Here the goal is to determine if a subject is healthy or diseased, or determine the disease state, based on a set of gene expression measurements. For example, one could run a microarray on a tissue or fluid sample from a subject then determine if the subject has a certain form of cancer. While this seems a straightforward manner of diagnosing patients, microarrays have yet to be routinely used in this fashion in clinical practice, but the hope is that one day this will be an effective method for diagnosing illness. Currently classification is used in a less formal way to identify genes that differ between patient groups (for example, Golub, 1999). While one can use methods for testing for differential expression, such methods don't directly use information on all genes simultaneously to detect differences in gene expression (although some methods do use some information from all genes to make a decision regarding each gene, such as SAM and the empirical Bayes approach discussed in Section 13.4). There are a number of good texts available that cover this material in greater depth, such as Hastie, Tibshirani, and Friedman (2001).

The data for the classification problem consists of a set of measurements x_i for all i arrays and an indicator y_i for each array indicating group membership. The problem is then to define a map $\mathcal{C}(x)$ (known as a the classifier) from the space whose domain is the space in which the x_i reside to the range where the y_i reside in some optimal fashion. Once one has determined $\mathcal{C}(x)$, given a new set of measurements for some subject x^{new}, one classifies this subject as $\mathcal{C}(x^{\text{new}})$. In the context of microarray analysis x_i is a set of gene expression measurements, so we will now assume that the x_i are p-dimensional vectors. Here p may be the number of genes on the microarray, but frequently only a subset of these genes are used for classification. The reason for using only a subset is that for any given tissue or fluid specimen only a subset of genes are actually expressed in the sample. Genes that aren't expressed are not useful for classification as the data for those genes is just random noise. In addition, sometimes researchers use a set of genes that is even smaller than the set of genes thought to be expressed in the tissue or fluid sample since genes whose level of expression doesn't differ between the groups will not be informative for constructing the classifier. Sets of techniques whose goal is to find a subset of genes that are most important for classification are known as filtering techniques (as in cluster analysis). For classification, filtering is frequently just done using tests for differential expression since genes that don't differ between classes won't discriminate between subjects from the classes.

Here we adopt the perspective that the goal of classification is to minimize the expected cost of misclassification (ECM). To simplify matters, suppose there are only 2 groups. Then our goal is partition $\mathbb{R}^p$ (i.e. p dimensional Euclidean space) into 2 regions, which we will denote R_1 and R_2 so that if $x^{\text{new}} \in R_1$ then we classify the subject from which x^{new} came as group 1 otherwise this subject is classified as in group 2. Let π_1 denote the proportion of the population from which the subjects are drawn that in truth belong to group 1 and let $\pi_2 = 1 - \pi_1$, the proportion of the population in the other group. Since the data x_i come from different groups and the idea is that the x_i differ between the 2 groups, we will suppose that the densities which give rise to the measurements differ between the 2 groups. We will denote these densities as $f_1(x)$ and $f_2(x)$. If $f(x)$ is the density for the data that marginalizes over group membership, then we have $f(x) = \pi_1 f_1(x) + \pi_2 f_2(x)$. Since there are only 2 groups, there are only 2 ways one can misclassify an observation: misclassify as group 1 or misclassify as group 2. Hence the expected cost of misclassification (ECM) is just the cost of misclassifying as group 1 (denoted c_1) times the probability of misclassifying as group 1 (denoted p_1) plus cost of misclassifying as group 2 (denoted c_2) times the probability of misclassifying as group 2 (denoted p_2). Next note that

$$
\begin{aligned}
p_1 &= P(x_i \in R_1 | \text{subject } i \text{ is in group 2}) P(\text{subject } i \text{ is in group 2}) \\
&= \pi_2 \int_{R_1} f_2(x)\, dx.
\end{aligned}
$$

Similarly we find that $p_2 = \pi_1 \int_{R_2} f_1(x)\, dx$, so that

$$
\begin{aligned}
\text{ECM} &= c_1 \pi_2 \int_{R_1} f_2(x)\, dx + c_2 \pi_1 \int_{R_2} f_1(x)\, dx \\
&= c_1 \pi_2 \left[1 - \int_{R_2} f_2(x)\, dx\right] + c_2 \pi_1 \int_{R_2} f_1(x)\, dx \\
&= c_1 \pi_2 + \int_{R_2} [c_2 \pi_1 f_1(x) - c_1 \pi_2 f_2(x)]\, dx.
\end{aligned}
$$

Note that the first term in the last expression does not depend on R_1 or R_2, hence given that our goal is to find the partition that minimizes the ECM, we can ignore this term. Thus the goal is to choose R_2 to include x values so that

$$
c_2 \pi_1 f_1(x) - c_1 \pi_2 f_2(x) < 0
$$

since then the integral will be negative. This result makes intuitive sense. For example if c_1 is huge then choose R_2 to be all of $\mathbb{R}^p$, i.e. if the cost of misclassifying a subject as being in group 1 is huge, then always classify a subject as being in group 2. Similarly, if π_1 is almost zero then always classify a subject as being in group 2.

The use of the ECM criterion is complicated in practice due to the presence of the costs and the population proportions. While sometimes the population proportions for a given disease are known, when constructing diagnostic criteria in practice one can't simply use these population proportions. The reason is, usually someone goes to a doctor because of some symptoms. The physician then might consider using some test for a disease which is based on his or her experience and knowledge in conjunction with the patient's personal information. For example, a patient might have a job that exposes him or her to toxins that may increase the risk for a certain type of cancer. If this person then developed a tumor after having this job for 25 years and went to a physician to have this checked, the relevant proportion for the decision making problem would not be the population proportion of those with this type of cancer, but some value which is higher. Specification of the expected cost of misclassification is also complicated since these could potentially include many things, ranging from lost income to even death. For these reasons, much of the research in classification "ignores" the costs associated with misclassification and the population proportions. We say ignores in quotes because in fact, much of the research can be considered as methods that minimize the ECM while assuming that $c_1 = c_2$ and $\pi_1 = \pi_2$. In practical medical decision making, these factors should be considered in combination with results from diagnostic criteria.

Notice that if we assume $c_1 = c_2$ and $\pi_1 = \pi_2$, then the set R_2 that minimizes the ECM is $\{x : f_2(x) > f_1(x)\}$. This makes intuitive sense in that we should assign cases to the groups from which it is more likely that they arose. Thus it would appear that we have converted our classification problem into the problem of nonparametric density estimation (for which there is a large literature; see, for example, Silverman, 1986; Scott, 1992; Efromovich, 1999), and while this is a useful approach in some contexts, this is not useful for the analysis of microarray data. The reason is that the number of genes for which we have data, p, typically greatly exceeds the number of samples, hence nonparametric density estimates will be highly variable and not useful in practice. A more pragmatic approach is to constrain the density estimate to some class (e.g. multivariate normal). In Section 15.3.1 we will consider this approach, but many methods have been proposed.

15.2 Cross-validation

Invariably, methods for classification depend on parameters that must be estimated from the data, so that we have $C_\theta(x)$ for some parameter (or collection of parameters), θ. One general method to approach this problem, called cross-validation, estimates θ by breaking the data into multiple groups, then using these groups to estimate θ and test the accuracy of the classifier. The data that is used to estimate θ is called the *training set* and the remainder is called the *test set*. The simplest example is called holdout validation. In this approach, some of the data is set aside simply for testing the accuracy of predictions made by the classifier. As such a method doesn't use all of the data for estimating θ it is clearly not the most efficient method for estimating any parameters that enter the classifier.

In K-fold cross-validation, the data set is first split into K subsamples. If one of the classes is very small, it may make sense to use stratified sampling so that each subsample has a sufficient number of observations from each class. Then, each of the subsamples is set aside and used as a test set while the remaining $K - 1$ subsamples are pooled together to estimate θ. If we denote this estimate $\hat{\theta}_k$, then we use $\mathcal{C}_{\hat{\theta}_k}(x)$ to classify each of the observations from the portion of the data that was set aside. Since the true class of these observations is known, we can then obtain an estimate of the probability of misclassification (this will be 2 proportions if there are 2 classes since there are 2 ways of misclassifying). This process of estimating θ is then repeated so that each subsample plays the role of the test set once. Then $\hat{\theta} = \frac{1}{K} \sum_k \hat{\theta}_k$ is typically used to estimate θ, although more sophisticated approaches are possible. For example, if there is an estimate of $\mathrm{Var}\, \theta_k$ we may downweight the more variable estimates in computing $\hat{\theta}$. Values of K are usually taken to be 5 or 10, depending on the size of the data set. Leave one out cross-validation is the case when K is taken equal to the sample size. While this would seem to provide the best estimate of θ since each of the $\hat{\theta}_k$ would have properties similar to an estimate that used all the data, it can be computationally intensive if the classification algorithm is slow or the data set is large.

15.3 Methods for classification

Here we describe some of the most popular methods that have been used for classification in microarray analysis.

15.3.1 Discriminate analysis

One way to minimize the ECM without fitting high dimensional nonparametric densities with sparse data sets is to assume a parametric specification for the densities that appear in the criterion. For example, if one supposes hat both densities are multivariate normal with the same covariance matrix but different means, then the classification rule one finds is equivalent to a method developed by R. A. Fisher in the 1930's known as linear discriminant analysis. If one supposes that the covariance matrices differ between the 2 classes then one obtains what is known as quadratic discriminant analysis. Both of these methods are used in the analysis of microarrays (see, for example, Dudoit, Fridlyand and Speed, 2002). In addition, diagonal linear discriminant analysis, which supposes the covariance matrices are not only the same for the 2 groups, but further assumes that these matrices are diagonal, has also been widely used. We will consider each of these methods in greater detail then compare these methods.

Following the previous approach to the minimum ECM method, we can characterize R_1 in quadratic discriminant analysis via

$$R_1 = \left\{ x : |\Sigma_1|^{-\frac{1}{2}} \exp\left\{ -\frac{1}{2}(x - \mu_1)'\Sigma_1^{-1}(x - \mu_1) \right\} \right.$$

$$\left. \times \quad |\Sigma_2|^{\frac{1}{2}} \exp\left\{ \frac{1}{2}(x - \mu_2)'\Sigma_2^{-1}(x - \mu_2) \right\} \geq \frac{c_1\pi_1}{c_2\pi_2} \right\}.$$

The function describing the boundary between R_1 and R_2 over $\mathbb{R}^p$ involves a quadratic in x and is thus not a connected set. For example, suppose $p = 1$, take $\frac{c_1\pi_1}{c_2\pi_2} = 1$, and consider Figure 15.1. As we noted above, when $c_1 = c_2$ and $\pi_1 = \pi_2$, then the set R_2 that minimizes the ECM is $\{x : f_2(x) > f_1(x)\}$, but when the variances differ we can clearly have R_1 be a non-connected set. When $p > 1$ R_1 can be very complicated. On the other hand, if we assume that $\Sigma_1 = \Sigma_2$, then the determinants in the expression that defines the boundary between the 2 regions cancel as does the quadratic term in the exponential, and if we set $\Sigma = \Sigma_1 = \Sigma_2$, then we find that

$$R_1 = \left\{ x : (\mu_1 - \mu_2)'\Sigma^{-1}x - \frac{1}{2}(\mu_1 - \mu_2)'\Sigma^{-1}(\mu_1 + \mu_2) \geq \log\left(\frac{c_1\pi_1}{c_2\pi_2}\right) \right\}.$$

This is the boundary that separates the 2 regions in linear discriminant analysis. Note that this boundary is described by an equation that is linear in x (hence the name), so that the boundary is just a hyperplane in $\mathcal{R}^p$. Such classification boundaries are easy to use and communicate with other researchers about. Finally, note that if we assume Σ is a diagonal matrix where the jth element on the diagonal is σ_j^2, and we suppose that the jth element of μ_i is μ_{ij} then

$$R_1 = \left\{ x : \sum_j (\mu_{1j} - \mu_{2j})x_j/\sigma_j^2 - \frac{1}{2}\sum_j(\mu_{1j} - \mu_{2j})(\mu_{1j} + \mu_{2j})/\sigma_j^2 \geq \log\left(\frac{c_1\pi_1}{c_2\pi_2}\right) \right\}.$$

Given the above characterizations of the set R_1 and given that linear discriminant analysis and diagonal linear discriminant analysis are special cases of quadratic discriminant analysis, it may appear that one should always use the most general model, quadratic discriminant analysis, in practice. While the set R_1 can be more complicated, in practice, to classify a given observation we need only substitute that value in the inequality to determine how we should classify the new observation. This is potentially odd in practice, for, suppose that $p = 1$ and consider Figure 15.1. In this case we could find that the rule for assigning a patient to diseased or healthy may have the difficult to interpret property that we say someone is healthy provided the outcome for the assay is such that the value is either too high or too low (most diagnostic criteria in practice have a single cut-off value or a range of healthy values). Another reason to not always use quadratic discriminant analysis is that the set R_1 depends on generally unknown population parameters (i.e. μ_i and Σ_i). In practice we must estimate these quantities, and since the simpler sets of assumptions lead to classification rules that involve fewer parameters, the simpler methods can outperform the more complex models in practice. Typically, if we let x_{ij} represent the p

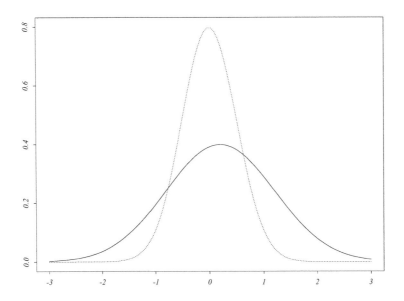

Figure 15.1 *An example of 2 univariate normal densities showing that the classification region for a group can be a disconnected set.*

vector of measurements for subject j in group i and we suppose that there are n_i observations in group i, then the μ_i are estimated by the mean in group i,

$$\overline{x}_i = \frac{1}{n_1} \sum_j x_{ij}$$

and Σ_i are estimated by the within group sample covaraince matrix,

$$S_i = \frac{1}{n_i - 1} \sum_j (x_{ij} - \overline{x}_i)(x_{ij} - \overline{x}_i)'.$$

In the context of linear discrimianant analysis, we use the pooled covariance matrix estimate of Σ,

$$S_p = \frac{n_1 - 1}{n_1 + n_2 - 2} S_1 + \frac{n_2 - 1}{n_1 + n_2 - 2} S_2.$$

Note that we will not be able to invert S_i unless $n_i > p$ or S_p unless $\max_i n_i > p$ so the inverted matrix that defines the boundary between R_1 and R_2 will not be usable. Since it is frequently the case that there are many more genes than arrays in a microarray study, one generally cannot use linear or quadratic discriminant analysis with the usual unbiased estimates given above, although one can still use diagonal linear discriminant analysis. Alternatively, one could use other estimates than the usual unbiased estimates of Σ_i that we have used here (see exercise 1 for a Bayesian approach).

There are a number of features of linear discriminant analysis that provide further insight into the problem of classification. Note that if we let $\theta' = (\mu_1 - \mu_2)'\Sigma^{-1}$, then we can write the region for group 1 in linear discriminant analysis as

$$R_1 = \left\{ x : \theta' x \geq \frac{1}{2}\theta'(\mu_1 + \mu_2) \right\}.$$

We can think of θ as a weighted estimate of the differences between the 2 groups (so genes which differ greatly in terms of their normalized difference give values to their corresponding element of θ with large absolute values), then a given gene gives evidence for class 1 if the value x_j exceeds the average across groups weighted by this same factor. Recall that linear discriminant analysis assumes both variances are the same so with normal densities we should just assign a case to the class with the closer mean, and here we are doing the same thing. In a certain sense, the classification decision is based on comparing each gene at a time then summing over all genes. For each gene, we compute how much we expect the level of gene expression to differ between the 2 classes.

For example if Σ is diagonal then $\theta_j = (\mu_{1j} - \mu_{2j})/\sigma_j^2$. This will be large if group 1 has a higher level of gene expression for this gene. Then multiply this value with x_j and the mean of x_j supposing there is no difference between the 2 groups, since, still assuming Σ is diagonal, this is just $\frac{1}{2}(\mu_{1j} + \mu_{2j})$. Also note that given θ is only determined up to a constant (i.e. $\theta' = c\theta$ gives the same classification boundary for any c), we can interpret the relative sizes of these coefficients as the importance of a particular gene for determining the classification boundary, provided the genes have all been standardized (i.e. divide the level of gene expression for each gene by the standard deviation of that gene across replicates). Note that the elments of θ have units given by the inverse of the units of measurement for a given element of x, hence if one has genes all obtained from the same microarray platform one could interpret all of these values as having the same unit of measurement and directly interpret the elments of θ since these would have units relating the measurement process. However if one wanted to use, say, blood pressure along with gene expression data for classification, one would want to standardize the variables first. Alternatively, one can use summaries such as $\frac{\theta}{\sqrt{\theta'\theta}}$ or θ/θ_i for arbitrary i to summarize the relative importance since both of these quantities are unit free.

15.3.2 Regression based approaches

Perhaps the most straightforward method for classification is to use logistic regression (as described in Section 13.3.2). In fact, one can just use linear regression since the cutoff rule for making a new prediction will involve comparing a predicted value for the new subject based on the regression coefficient estimates, and if this prediction is outside the interval (0,1) this will not really matter. In this approach one would use the class membership as the response variable and find a set of predictors (i.e. gene expression levels for many genes) that determines as much of the variation in the response variable as possible. The primary obstacle to a straightforward implementation of this approach is the issue of model selection. Usually there are many more genes than subjects, hence one can't fit a model that includes all of the potential predictors since there will not be a unique value for the parameter estimate. One could try fitting many univariate models and construct a model using as many variables as possible (starting with those that are the strongist predictors of class membership) but there will typically still be too many genes to allow consistent estimation of all parameters. More recent attempts to use a straightforward regression based approach have added penalty terms to the usual log-likelihood, or equivalently from a Bayesian perspective, have used a prior on the regression coefficients. The penalty (or prior) has the effect of shrinking the coefficients to zero but allows one to use more genes in the regression equation. In fact, if one uses a penalty term that depends on the absolute value of the regression coefficients, i.e. minimize

$$\sum_{i=1}^{n}(y_i - x_i'\beta)^2 + \lambda \sum_{j=1}^{p}|\beta_j|$$

then many of the regression coefficients obtained from minimizing the penalized sum of squares will be identically equal to zero, thereby providing model selection and smoothing of the parameter estimates simultaneously (this algorithm has been named "the lasso"; Tibshirani, 1996). These algorithms depend on a parameter, λ, that governs how much the coefficients should be shrunk towards zero: such parameters are usually estimated using cross-validation.

15.3.3 Regression trees

A regression tree is a set of nodes and edges that define a classifier. At the top node, the data is split into 2 groups, and at each lower node, the data is again split into 2 groups. After a certain number of these cascading bifurcations, the tree structure ends and all items at that point are assigned to a class. While details of implementation vary, generally a very large tree is constructed with the goal of choosing the split points so as to minimize the heterogeneity within a node. After the large tree is constructed, it is pruned to obtain a smaller tree and cross validation is used to estimate an error rate. For microarray analysis, the splits are defined by the level of expression for some gene. These methods define regions in the space of gene expression measurements that have nonlinear boundaries. For example, suppose there are just 2 genes. Then the region that results in classification to class 1 could look like 2 squares that touch only at one corner, hence this is considered to be a nonlinear method. The details of using these methods are quite complex, but software is widely available (one popular version is called CART; see Breiman et al., 1984). While the CART algorithm frequently performs well in practice, it is not based on any explicit probability model and so can't provide statistical summaries of the fit of the method. In addition, there is a Bayesian versions (BART) that performs well in practice and does have an explicit probability model as the basis of the algorithm thus providing statistical summaries in terms of posterior probabilities (see Chipman, George, and McCulloch, 2006).

15.3.4 Weighted voting

Recall from the discussion of linear discriminant analysis how we can view the classifier as based on allowing each gene to play a role in determining the relevance for including the information from that gene in the classifier by computing a standardized distance between the 2 means. Then one compares the value of x to the pooled sample mean in terms of how relevant that gene is for discriminating between the groups. Weighted voting (Golub et al., 1999) takes its name from an algorithm that implements this idea in a slightly different fashion. This was one of the first algorithms applied in microarray analysis and the important difference from methods we've discussed thus far in that the method does not predict the class of all cases. Instead such observations are said to be marginal, and no prediction is made.

Recall from the discussion of diagonal linear discriminant analysis that we can write the estimated region R_1 as

$$
\begin{aligned}
R_1 &= \left\{ x : \sum_j \frac{(x_j - \overline{x}_{2j})^2}{\hat{\sigma}_j^2} \geq \sum_j \frac{(x_j - \overline{x}_{1j})^2}{\hat{\sigma}_j^2} \right\} \\
&= \left\{ x : \sum_j \frac{(\overline{x}_{1j} - \overline{x}_{2j})}{\hat{\sigma}_j^2} (x_j - \frac{(\overline{x}_{1j} + \overline{x}_{2j})}{2}) \geq 0 \right\}.
\end{aligned}
$$

Similarly, in weighting voting, if we let $\hat{\sigma}_{ij}^2$ represent the usual unbiased estimate of the variance in class i for the jth gene, the region R_1 is defined as

$$
R_1 = \left\{ x : \sum_j \frac{(\overline{x}_{1j} - \overline{x}_{2j})}{\hat{\sigma}_{1j} + \hat{\sigma}_{2j}} (x_j - \frac{(\overline{x}_{1j} + \overline{x}_{2j})}{2}) \geq 0 \right\}.
$$

While this is similar to the region used in diagonal linear discriminant analysis, there are 2 differeces: the variance estimate does not pool across classes and the standard deviation is used in place of the variance. The use of the standard deviation in place of the variance implies that R_1 depends on the units used, in contrast to the region defined by diagonal linear discriminant analysis. If we define

$$
v_j = \frac{(\overline{x}_{1j} - \overline{x}_{2j})}{\hat{\sigma}_{1j} + \hat{\sigma}_{2j}} \left(x_j - \frac{(\overline{x}_{1j} + \overline{x}_{2j})}{2} \right),
$$

then we can think of each gene as "voting" for each class since we sum the v_j's to determine R_1, where each vote has been weighted by the sum of the within class standard deviations. If we further define $v_1 = \sum_j \max(v_j, 0)$ and $v_2 = \sum_j \max(-v_j, 0)$, then v_1 indicates how strongly the data votes for class 1 (there is an analogous interpretation for v_2). The measure of the prediction strength is then defined to be $PS = \frac{|v_1 - v_2|}{v_1 + v_2}$, and a prediction is only made for a new observation if $PS > 0.3$. This algorithm has been found to work well in practice.

15.3.5 Nearest neighbor classifiers

Nearest neighbor classification is another popular classification method that requires only a way to compute a measure of similarity between 2 items that are being classified. An object is classified as belonging to the class determined by the majority of its neighbors. The number of neighbors used to make this decision is typically determined by cross-validation. So, for each possible number of neighbors, one determines the error rate (usually treating the 2 possible misclassifications when there are only 2 classes as being equal), then selects the number of neighbors to use by minimizing this error rate.

15.3.6 Support vector machines

Support vector machines (SVM) are a collection of methods for classification that are widely used and typically have performance that is competitive with other methods (see the brief tutorial by Burges, 1998; or the texts Vapnik, 1982; Vapnik, 1995 for greater detail, and Brown et al., 2000, for an early microarray application). The basic idea is that, although the data may not be well separated in the space in which they reside, perhaps by embedding them in a higher dimensional space we can separate them easily. The approach is designed for classification for only 2 groups, although there are a number of possible ways to extend the method to more than 2 classes.

We suppose we have data (x_i, y_i) where x_i are p dimensional vectors and y_i is equal to 1 or -1 indicating to which of the classes observation i belongs. Suppose we have a mapping ϕ from $\mathbb{R}^p$ to some vector space F called the feature space. Although we can think of F as some higher dimensional Euclidean space, it is potentially more general. For example, if $p = 1$ and feature space is 2-dimensional Euclidean space, then we can use the map $\phi(x) = (x, \alpha)'$ for some α. The goal is find a mapping that separates the 2 groups in feature space. To measure the amount of separation between 2 groups of points, we will suppose that we can find a map to feature space so that there exists a hyperplane that goes between the 2 groups. For our simple example where $p = 1$ and feature space is a 2-dimensional Euclidean space, if we let B represent some subset of the real line (not necessarily a connected set) such that $x_i \in B$ for all i where $y_i = 1$ and $x_i \notin B$ for all i where $y_i = -1$, then define ϕ so that $\phi(x) = (x, \alpha_1)$ for $x \in B$ and $\phi(x) = (x, -\alpha_2)$ for $x \notin B$ for some $\alpha_1 > 0$ and $\alpha_2 > 0$ then any line $x_2 = c$ for $|c| < \min |\alpha_i|$ will separate the 2 groups in feature space (note that this generalizes to the more general case $p > 1$ quite easily).

In practice one does not specify the mapping to feature space or even feature space itself. Rather, one specifies what is called the *kernel function*. To understand why, we define the notion of the margin. Suppose we can find a hyperplane that separates the 2 classes in feature space. As the example above indicates, in general, there are then an infinite number of hyperplanes that separate the 2 classes. We will parameterize our hyperplanes with (w, b) where w is a vector normal to the hyperplane and $w'x + b = 0$ for all x that reside on this hyperplane. Let d_+ be the shortest distance from the

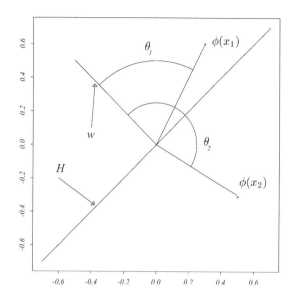

Figure 15.2 *A plot of 2 points ($\phi(x_1)$ and $\phi(x_2)$) that have been mapped to feature space (here just the plane). A separating hyperplane is represented by H and a vector normal to H is shown and labeled w. Note that* $\cos\theta_1 > 0$ *while* $\cos\theta_2 < 0$.

separating hyperplane to a member of the set of $\phi(x_i)$ where $y_i = 1$ and define d_- analogously for the case where $y_i = -1$. The margin is defined to be $d_+ + d_-$ and represents the distance between the 2 groups after mapping the observations to feature space. For an arbitrary mapping, our goal is then to find the hyperplane that maximizes the margin assuming such a hyperplane exists. For our simple 2-dimensional Euclidean space example given above, we should clearly take $\alpha \to \infty$ and the margin is unbounded, although use of different mappings will produce margins which are bounded.

To maximize the margin in feature space assuming that the groups are separated in feature space is a constrained maximization problem. For our hyperplane to separate the groups in feature space, we require that

$$\phi(x_i)'w + b \geq 0 \text{ for } y_i = 1$$
$$\phi(x_i)'w + b \leq 0 \text{ for } y_i = -1. \tag{15.1}$$

This is because (as illustrated in Figure 15.2), for 2 vectors x and w in Euclidean space, if $\|x\|$ represents the norm (i.e. length, or distance of the point x to the origin) of the vector x, then $x'w = \|x\|\|w\|\cos\theta$ where θ is the angle between the 2 vectors (recall that $\cos$ changes sign at $\pi/2$ and $3\pi/2$). We can combine the n conditions in the pair of equations 15.1 into the more simple condition

$$y_i(\phi(x_i)'w + b) \geq 0$$

for all $i = 1, \ldots, n$. Our goal is to then maximize the margin subject to these constraints.

Consider the set of points where $\phi(x_i)'w + b = \epsilon_i$ where $y_i = 1$, ϵ_i is positive and the smallest of these ϵ_i that is greater than zero. Consider also the analogous set of points in the group where $y_i = -1$, i.e. those closest to the separating hyperplane. Such points lie on the hyperplane $w'\phi(x_i) + b = \epsilon_i$ and $w'\phi(x_j) + b = -\epsilon_j$, or $w'\phi(x_i) + b - \epsilon_i = 0$ and $w'\phi(x_j) + b + \epsilon_j = 0$. Next note that the perpendicular distance from the hyperplane (w, b) to the origin is just $|b|/\|w\|$ (this is shown by finding the values of x that minimize the distance of a hyperplane to the origin, then substituting those values of x into the expression for the distance of a hyperplane to the origin). This implies that $d_+ = |b - \epsilon_i|/\|w\|$ and $d_- = |b + \epsilon_j|/\|w\|$, so that provided $|b| < 1$ (which we assume here) $d_+ + d_- = 2/\|w\|$. Hence the hyperplane that maximizes the margin is found by minimizing $1/\|w\|$ subject to the constraints in equation 15.1.

We can then define the Lagrangian for inequality constrained maximization as

$$\mathcal{L}(w, b, \alpha_1, \ldots, \alpha_n) = \frac{1}{2}\|w\|^2 - \sum_i \alpha_i y_i(\phi(x_i)'w + b) + \sum_i \alpha_i \quad (15.2)$$

where $\alpha_i \geq 0$ are a set of Lagrange multipliers (assumed positive due to the presence of inequality constraints). First, we differentiate $\mathcal{L}(w, b, \alpha_1, \ldots, \alpha_n)$ with respect to w to get the set of equations:

$$w = \sum_j \alpha_j y_j \phi(x_j).$$

Some authors suggest also differentiating this with respect to b to get the equation

$$\sum_i \alpha_i y_i = 0,$$

however $\mathcal{L}$ is unbounded as a function of b (since it is linear) so we will just set $b = 0$ since this has the same effect. If we then substitute the expression for w back into equation 15.2 and use the fact that $b = 0$ we get the following function that we must maximize to find values for the Lagrange multipliers

$$\mathcal{L}'(\alpha_1, \ldots, \alpha_n) = \sum_i \alpha_i - \frac{1}{2}\sum_{ij} \alpha_i \alpha_j y_i y_j \phi(x_i)'\phi(x_j).$$

This is a quadratic programming problem and the solution to such problems is well understood (see, for example, the texts of Mangarasian, 1969; McCormick, 1983; Fletcher, 1987; Moré and Wright, 1993). One property of such problems is that some of the α_i will be zero, and some non-zero. The i such that $\alpha_i > 0$ correspond to x_i that define the separating hyperplane and are referred to as support vectors. Once we have the hyperplane we can use it by just computing

$$w'\phi(x^{\text{new}}) + b = \sum_j \alpha_j y_j \phi(x_j)'\phi(x^{\text{new}}),$$

and if this is greater than zero we classify it as belonging to the $y = 1$ group, otherwise classify x^{new} as belonging to the other group.

Note that we can find the optimal set of α_i just using inner products in feature space. Also note that once we have the optimal set of α_i, w is just a linear combination of these and $\phi(x_i)$, so that when using the decision procedure we just compute inner products in feature space after mapping both the data set and x^{new} to feature space. Thus we don't really need to specify ϕ and feature space, we need only define $K(x,y) = \phi(x)'\phi(y)$ for some K that depends on 2 vectors in the space in which our data resides. We call K the kernel function, and by specifying a kernel function we can use support vector machines without specifying feature space or the mapping to feature space. Since K plays the role of an inner product after to mapping to some space, there are constraints on K that must be satisfied. In particular, we require that K is an allowed kernel function, and this is true due to a result known as Mercer's theorem (see, for example, Porter and Stirling, 1990) if and only if

$$\int K(x,y)g(x)g(y)\,dx\,dy \geq 0$$

for any g so that

$$\int |g(x)|^2\,dx < \infty$$

(in this case K is said to be a positive semi-definite function). For example,

$$K(x,y) = (x'y)^p,$$
$$K(x,y) = e^{-\|x-y\|^2/\sigma}$$

and

$$K(x,y) = (x'y + 1)^p$$

are allowed and commonly used.

While we know we can find a separating hyperplane if we choose the map wisely (consider the simple example where $p = 1$), if we instead specify the kernel function there is no guarantee that the points will be separable in the implicitly defined feature space using the implicitly defined map. In this case there is no solution to our inequality constrained problem since there is no solution that satisfies the set of constraints. To overcome this problem, we can introduce a set of slack constraints, ξ_i for $i = 1, \ldots, n$ and modify the set of constraints presented in equations 15.1 to include these variables as follows

$$\phi(x_i)'w + b \;\geq\; -\xi_i \text{ for } y_i = 1$$
$$\phi(x_i)'w + b \;\leq\; \xi_i \text{ for } y_i = -1. \qquad (15.3)$$

We account for these errors by introducing a penalty term to our Lagrangian that penalizes the existence of such mistakes. An easy way to accomplish this is to replace $\|w\|^2/2$ with $\|w\|^2/2 + C\sum_i \xi_i$ where a larger value of C is associated with making more errors on the data set used to estimate the classifier. This parameter is either specified by the user or estimated via cross-validation. Thus by specifying a kernel and a value of C one can obtain the classifier using standard optimization methods.

For our simple example where feature space is 2-dimensional Euclidean space and we use the mapping $\phi(x) = (x, \alpha_1)$ for $x \in B$ and $\phi(x) = (x, -\alpha_2)$ for $x \notin B$ for some fixed, finite α_1 and α_2, then the hyperplane that maximizes the margin is described by $w = w_0(0, 1)'$ where $w_0 = (\alpha_1 - \alpha_2)/2$ and $b = -w_0$. Also note that we will classify a subject as belonging to the group with $y_i = 1$ if $x^{new} \in B$ and classify x^{new} as belonging to the other group otherwise. The use of this ϕ as a mapping function is very similar to nearest neighbor classification since x^{new} is classified as belonging to a group depending on how close the observation is to the data used to construct the classifier. Finally, notice that the kernel that corresponds to this choice of ϕ is

$$K(x, y) = \begin{cases} x'y + \alpha_1^2 & \text{if } x \in B \ y \in B \\ x'y + \alpha_2^2 & \text{if } x \notin B \ y \notin B \\ x'y - \alpha_1\alpha_2 & \text{if } x \notin B \ y \in B \text{ or } x \in B \ y \notin B \end{cases}$$

which is very similar to the commonly used kernel $K(x, y) = x'y$ especially as α_1 and α_2 approach zero.

15.4 Aggregating classifiers

Frequently highly non-linear methods for classification give rise to unstable classifiers in the sense that slight changes in the data give large changes in the classifier. For this reason a number of approaches have been developed that involve the idea of slightly altering, or perturbing, the data and recomputing the classifier. One then averages over all these slightly varying classifiers to get the classifier which is then used. These methods work well in practice, and the 2 most commonly used methods are known as boosting and bagging.

15.4.1 Bagging

Bagging uses bootstrap samples of the data matrix to develop the classifier. One starts by randomly drawing a sample with replacement of size n from the set of (x_i, y_i) pairs. With this sample, one then develops the classification rule. Note that in this approach, whenever a sample is drawn, some of the pairs (x_i, y_i) will not be drawn, hence one can use these samples to estimate the misclassification rates. One problem with this approach is that the sampled data sets will have duplicate sets of x_i whereas real gene expression data sets never have this feature. In general, two approaches have been developed to avoid this situation. One approach is to use the parametric bootstrap to draw the samples. In the parametric bootstrap, one assumes some probability distribution for the observed data, such as a mixture multivariate normal distribution

$$p(y = 0)N(x|\mu_1, \Sigma_1) + p(y = 1)N(x|\mu_2, \Sigma_2))$$

where μ_1 and Σ_1 are the mean and covariance of x when $y = 0$ and μ_2 and Σ_2 are the mean and covariance of x when $y = 1$. One then estimates the parameters in this model then draws samples from the resulting probability distributions. Here one could just use the MLEs for all the parameters. That is, one could use the sample proportions to estimate $p(y = 0)$ and $p(y = 1)$ and use the mean and variance of the x_i so that $y_i = 0$ to estimate μ_1 and Σ_1 (estimates of μ_2 and Σ_2 can be found similarly). However, in order to sample from this distribution, one needs to be able to invert the covariance matrix, and here no such inverse will exist since $p > n$ in typical microarray studies. One could attempt to overcome this difficulty by supposing that Σ has some structure so that its MLE has an inverse, however it is preferable to use other methods that don't require such strong assumptions. One such approach is the use of convex pseudo-data. In this approach we select a pair of observations from the data set, (x_i, y_i) and (x_j, y_j), then combines these to form the observed pair

$$(x_i^{\text{simulated}}, y) = \left(ux_i + (1 - u)x_j, y_j\right),$$

where u is a uniform random variable on the interval $(0, d)$ for $d < 1$. A drawback of this approach is that one can potentially induce strong associations between distinct $x_i^{\text{simulated}}$ since they may be based on a common x_i, and this is also not a feature we observe in datasets. For more on bagging consult Breiman (1998a) and Breiman (1998b).

15.4.2 Boosting

In boosting, the data is adaptively resampled as the algorithm proceeds (see Valiant (1984) or Kearns and Vazirani (1994) for a more thorough introduction). While there are many variations on the basic idea, we will focus on a method known as AdaBoost (Freund and Schapire, 1997). Central to the method is the set of weights, $p_i^{(j)}$, at the jth iteration for sampling observation i for $i = 1, \ldots, n$ from the data set. Given $p_i^{(0)} = \frac{1}{n}$ for all i, the algorithm proceeds through the following steps (this is the jth iteration):

1. use $p_i^{(j)}$ to draw a sample with replacement of size n from the original data
2. use the observations sampled in step 1 to develop the classifier, call it $C_j(x)$
3. classify each observation in the data set using the classifier developed in step 2 and let $d_i = 0$ if observation i is classified incorrectly and set $d_i = 1$ otherwise
4. let $\epsilon_j = \sum_i p_i^{(j)} d_i$ and $\beta_j = \frac{1-\epsilon_j}{\epsilon_j}$ then set $p_i^{(j+1)} = \frac{p_i^{(j)} \beta_j^{d_i}}{\sum_i p_i^{(j)} \beta_j^{\frac{1}{2}-d_i}}.$

Then after B steps all the classifiers are aggregated by using the classifier

$$C(x) = \text{argmax}_k \sum_j \log \beta_j 1_{\{C_j(x)=k\}}.$$

Note that the algorithm tends to sample those observations that the classifier has a hard time of classifying, so that in a sense the method works hard at predicting the cases that are difficult to predict. Nonetheless, the algorithm is resistant to overfitting, although the reasons for this are controversial (see Freidman, Hastie and Tibshirani, 2000).

15.4.3 Random forests

One of the best performing classification methods currently is called random forests (Hastie, Tibshirani, and Friedman, 2001). This method generates many trees and uses these for classification. The idea is, given x^{new}, we use each of the trees to predict the class based on this collection of variables (so each tree gets to vote for the class of x^{new}). The class for x^{new} is determined by the majority of the votes.

Each of the trees in the forest is constructed (or "grown") in the same fashion. If there are n samples in the training set, for each tree we sample n observations from the training data set with replacement. Supposing there are p variables (i.e. genes) used to construct the trees, select a value m so that $m < p$. As the trees are grown, at each node a sample of m variables is selected and used to determine the split. All trees are grown as large as possible and no pruning occurs at the end of the tree construction phase. The parameter m can have a large effect on the performance of the algorithm. If m is too small then the trees tend to be very different, but as each only uses a small portion of the data, they also tend to perform poorly. Hence this parameter must be selected based on the data set used for classification. Since the data used for each tree is selected with replacement, the excluded cases can be used to obtain an unbiased error rate for each tree. These error rates can then be used to find appropriate values for m. The algorithm also determines the importance of a variable using the excluded cases. This is done by determining how each tree votes for each of the excluded cases and comparing this to how each tree votes for the excluded cases where the values for variable x_i are permuted. Intuitively, if permuting the values of a variable has little impact on the classification, then that variable is not very useful for classification. If the number of variables used to classify is very large, the algorithm can be run using all of the data once, then selecting variables to use in the classifier on the basis of these variable importance measures.

15.5 Evaluating performance of a classifier

The evaluation of the performance of a diagnostic technique is a problem that has been widely studied. Using a microarray to classify a subject is in many ways the same as using blood pressure to identify patients at high risk for cardiovascular events. The usual methods for assessing the usefulness of a diagnostic method have focused on the sensitivity and the specificity of method. The sensitivity is the probability that the test identifies someone with a disorder given that the individual has the disorder and the specificity is the probability that the test is negative conditional on someone not having the disorder. We want both of these probabilities to be high. Consider some variable that we think predicts if someone has a disorder where higher values of the variable are indicative of a greater likelihood that one has the disorder. If the decision is made that someone is classified as being affected if the value of this variable exceeds some cut-off criterion, then we have a diagnostic method with an implied sensitivity and specificity. If we take the cutoff value to be very low, then almost everyone will be found to be affected. This implies that the specificity will be very low and the sensitivity will be quite high (simply because the cutoff value is so low that no one will be classied as negative, so that the probability that the test is negative will be zero no matter what). If we take the cut off to be very high, then the specificity will increase and will eventually rise to 1 as we increase the cutoff level to values that are never observed. Conversely, as the cutoff level drops, fewer patients are classified as positive and the sensitivity drops, eventually to 0. For intermediate values we will have pairs of sensitivity and specificity values. This collection of points is frequently graphically displayed in a plot that has 1-specificity on the x axis and sensitivity on the y axis. Such a plot is called an receiver operating characteristic (ROC) curve. If the variable is strongly predictive of disease status then the ROC curve should rise above the $y = x$ and have an area under the curve near 1. On the other hand, if the variable used for diagnostics is independent of disease status then the ROC curve will be a $y = x$ line (see exercise 3). A plot of the ROC curve and computation of the area under the curve are commonly used summaries of the potential of a variable's use in diagnostics. If we have some variable that we want to use for diagnostic purposes it is straightforward to naively estimate the ROC curve using sample proportions, but there are sophisticated methods that smooth the proportions (see Green and Swets, 1966; Dodd and Pepe, 2003, for further information).

While it seems straightforward to establish the usefulness of a classifier using an ROC curve, a common problem in classification is that the researcher would like to know how well a classifier will perform, but only has the data set used to develop the classifier. If one simply uses the same data for both purposes then one will overstate confidence in the results. For this reason, cross-validation is frequently used when estimating a ROC curve.

15.6 Exercises

1. Describe how one could use an informative Wishart prior for Σ and describe R_1.

2. Fisher's approach to classification: find the classifier by constructing a linear combination of the data for each item that maximally separates the 2 groups.

3. Show that the ROC curve is a $y = x$ line if the variable used to conduct diagnostics is independent of whether or not one has the disease.

4. In the context of quadratic discriminant analysis, show the shape of ROC curve if you use the linear classification approach (i.e. as cut off value goes up then disease chances go up) when you actually have a disjoint region (i.e. variance is larger for the disease group). Can the curve be non-concave?

5. In a paper by Campo Dell'Orto et al. (2007), the authors compare 3 RNA preparation systems in terms of the ability of the system to allow one to distinguish between different subtypes of leukemia. The data is available through the gene expression omnibus (accession number GSE7757).

   (a) Use diagonal linear discriminant analysis for a fixed value of $C = \log\left(\frac{c_1 \pi_1}{c_2 \pi_2}\right)$ to construct a classifier using the data from each preparation system separately. Construct a ROC curve using this classifier by varying C.

(b) Repeat the previous part, but use a SVM to construct the classifier. What parameter must we vary to construct a ROC curve?

(c) Construct a test statistic that can be used to find sets of genes that differ among the subtypes of leukemia. Repeat the previous 2 parts, but now only use genes that differ among the varieties of leukemia you find using your test statistic while controlling the FDR at a reasonable level.

(d) Which preparation system works the best? Does it depend on your classification method? Explain.

6. In a paper by Pohlers et al. (2007), the authors used Affymetrix microarrays to distinguish between tissue samples from 2 different varieties of arthritis. The data is available through the gene expression omnibus (accession number GSE7669) in the form of a signal value produced by the MAS 5.0 algorithm (the probe level summaries are also available in the form of a .CEL file).

(a) Using just the spots that are found to be present using the Affymetrix presence call, construct a classifier using k nearest neighbors. Use leave one out cross validation to determine the number of neighbors one should use.

(b) Use the lasso to construct a classifier and compare to your previous findings.

(c) Use the lasso to construct a classifier using all of the genes on the array. How does this compare to your previous results?

References

G.R. Abecasis, S.S. Cherny, W.O. Cookson, and L.R. Cardon. GRR: Graphical representation of relationship errors. *Bioinformatics*, 17:742–743, 2001.

G.R. Abecsis, S.S. Cherny, W.O. Cookson, and L.R. Cardon. Merlin-rapid analysis of dense genetic maps using sparse gene tree flows. *Nature Genetics*, 30:97–101, 2002.

D.B. Allison. Transmission-disequilibrium test for quantitative traits. *American Journal of Human Genetics*, 60:676–690, 1997.

L. Almasy and J. Blangero. Multipoint quantitative-trait linkage analysis in general pedigrees. *American Journal of Human Genetics*, 62:1198–1211, 1998.

O. Alter, P.O. Brown, and D. Botstein. Singular value decomposition for genome-wide expression data processing and modeling. *Proceedings of the National Academy of the Sciences USA*, 97:10101–10106, 2000.

S.F. Altschul and W. Gish. Local alignment statistics. *Methods in Enzymology*, 266:460–480, 1996.

S.F. Altschul, W. Gish, W. Miller, E.W. Myers, and D.J. Lipman. Basic local alignment search tool. *Journal of Molecular Biology*, 215:403–410, 1990.

S.F. Altschul, T.L. Madden, A.A. Schaffer, et al. Gapped BLAST and PSI-BLAST: A new generation of protein database search programs. *Nucleic Acids Research*, 25:3389–3402, 1997.

P. Baldi, Y. Chauvin, T. Hunkapiller, and M.A. McClure. Hidden Markov models of biological primary sequence information. *Proceedings of the National Academy of Sciences USA*, 91:1059–1063, 1994.

P. Baldi and A.D. Long. A Bayesian framework for the analysis of microarray expression data: Regularized *t*-test and statistical inference of gene changes. *Bioinformatics*, 17:509–519, 2001.

J. Banfield and A. Raftery. Model-based Gaussian and non-Gaussian clustering. *Biometrics*, 49:803–822, 1993.

C. Barreau, T. Watrin, H. Beverley Osborne, and L. Paillard. Protein expression is increased by a class III AU-rich element and tethered CUGBP1. *Biochemical Biophysical Research Communications*, 347:723–730, 2006.

L.E. Baum. An equality and associated maximization technique in statistical estimation for probabilistic functions of Markov processes. *Inequalities*, 3:1–8, 1972.

A.D. Baxevanis and ed.s Ouellette, B.F.F. *Bioinformatics*. Wiley, New York, 2001.

J.M. Bernardo, M.J. Bayarri, J.O. Berger et. al., Eds. *Bayesian Statistics 7: Proceedings of the 7th Valencia International Meeting*. Clarendon Press, Oxford, 2003.

J. Blangerao, J.T. Williams, and L. Almasy. Quantitative trait locus mapping using human pedigrees. *Human Biology*, 72:35–62, 2000.

M. Boehnke, K. Lange, and D.R. Cox. Statistical methods for multipoint radiation hybrid mapping. *American Journal of Human Genetics*, 49:1174–1188, 1991.

B.M. Bolstad, R.A. Irizarry, M. Astrand, and T.P. Speed. A comparison of normalization methods for high density oligonucleotide array data based on bias and variance. *Bioinformatics*, 19:185–193, 2003.

D. Botstein, R.L. White, M.H. Skolnick, and R.W. Davies. Construction of a genetic linkage map in man using restriction fragment length polymorphisms. *American Journal of Human Genetics*, 32:314–331, 1980.

D. Bowtell and J. Sambrook. *DNA Microarrays*. Cold Spring Harbor Press, Cold Spring Harbor, NY, 2003.

L. Breiman. Arcing classifiers. *The Annals of Statistics*, 26:801–824, 1998.

L. Breiman. Using convex pseudodata to increase prediction accuracy. Technical Report 513, University of California, Berkeley, Department of Statistics, 1998.

L. Breiman, J. H. Friedman, R. Olshen, and C. J. Stone. *Classification and Regression Trees*. Wadsworth, Belmont, CA, 1984.

M.P.S. Brown, W. Noble Grundy, D. Lin, et al. Knowledge-based analysis of microarray gene expression data by using support vector machines. *Proceedings of the National Academy of the Sciences USA*, 97:262–267, 2000.

C. Burge and S. Karlin. Prediction of complete gene structures in human genomic DNA. *Journal of Molecular Biology*, 268:78–94, 1997.

C.J.C. Burges. A tutorial on support vector machines for pattern recognition. *Data Mining and Knowledge Discovery*, 2:121–167, 1998.

M. Burset and R. Guigó. Evaluation of gene structure prediction programs. *Genomics*, 34:353–367, 1996.

A.J. Butte and I.S. Kohane. Mutual information relevance networks: Functional genomic clustering using pairwise entropy measurements. In *Pacific Symposium on Biocomputing*, pages 418–429, 2000.

R.B. Calinski and J. Harabasz. A dendrite method for cluster analysis. *Communications in Statistics*, 3:1–27, 1974.

M. Campo Dell'Orto, A. Zangrando, L. Trentin, R. Li, et al. New data on robustness of gene expression signatures in leukemia: comparison of three distinct total rna preparation procedures. *BMC Genomics*, 22:188, 2007.

L.R. Cardon and G.D. Stormo. An expectation maximization algorithm for identifying protein binding sites with variable gaps from unaligned DNA fragments. *Journal of Molecular Biology*, 223:159–170, 1992.

B.P. Carlin and T.A. Louis. *Bayesian Methods for Data Analysis*. CRC Press, Boca Raton, FL, 2009.

C.K. Carter and R. Kohn. On Gibbs sampling for state space models. *Biometrika*, 81:541–553, 1994.

C.Y. Chen, R. Gherzi, S.E. Ong, E.L. Chan, R. Raijmakers, and G.J. Pruijn et al. AU binding proteins recruit the exosome to degrade ARE-containing mRNAs. *Cell*, 107:451–464, 2001.

D. Cheon, C. Chae, and Y. Lee. Detection of nucleic acids of porcine reproductive and respiratory syndrome virus in the lungs of naturally infected piglets as determined by *in-situ* hybridization. *Journal of Comparative Pathology*, 117:157–163, 1997.

V.G. Cheung, M. Morley, F. Aguilar, A. Massimi, R. Kucherlapati, and G. Childs. Making and reading microarrays. *Nature Genetics Supplement*, 21:15–19, 1998.

H.A. Chipman, E.I. George, and R.E. McCulloch. Bart: Bayesian additive regression trees. Technical report, University of Pennsylvania, 2006.

A.G. Comuzzie, J.T. Williams, L.J. Martin, and J. Blangero. Searching for genes underlying normal variation in human adiposity. *Journal of Molecular Medicine*, 79:57–70, 2001.

The Gene Ontology Consortium. Gene ontology: tool for the unification of biology. *Nature*

Genetics, 25:25–29, 2000.

E. Crowley, K. Roeder, and M. Bina. A statistical model for locating regulatory regions in genomic DNA. *Journal of Molecular Biology*, 268:8–14, 1997.

S.E. Daniels, S. Bhattacharrya, A. James, N.I. Leaves, A. Young, et al. A genome-wide search for quantitative trait loci underlying asthma. *Nature*, 383:247–250, 1996.

B. Devlin and K. Roeder. Genomic control for association studies. *Biometrics*, 55:997–1004, 1999.

L.E. Dodd and M.S. Pepe. Partial AUC estimation and regression. *Biometrics*, 59:614–623, 2003.

X. Duan, H. Nauwynck, and M. Pensaert. Virus quantification and identification of cellular targets in the lungs and lymphoid tissues of pigs at different time intervals after inoculation with porcine reproductive and respiratory syndrome virus (PRRSV). *Veterinary Microbiology*, 56:9–19, 1997.

J. Dudoit, J. Fridlyand, and T.P. Speed. Comparison of discrimination methods for the classification of tumors using gene expression data. *Journal of the American Statistical Association*, 97:77–87, 2002.

S. Dudoit, Y. Yang, M. Callow, and T.P. Speed. Statistical methods for identifying differentially expressed genes in replicated cdna microarray experiments. *Statistica Sinica*, 12:111–139, 2002.

D.J. Duggan, M. Bittner, Y. Chen, P. Meltzer, and J.M. Trent. Expression profiling using cDNA microarrays. *Nature Genetics Supplement*, 21:10–13, 1999.

R. Durbin, S. Eddy, A. Krogh, and G. Mitchison. *Biological Sequence Analysis*. Cambridge University Press, UK, 1998.

S. Efromovich. *Nonparametric Curve Estimation: Methods, Theory and Applications*. Springer, New York, 1999.

B. Efron, R. Tibshirani, J.D. Storey, and V. Tusher. Empirical bayes analysis of a microarray experiment. *Journal of the American Statistical Association*, 96:1151–1160, 2001.

M.B. Eisen, P.T. Spellman, P.O. Brown, and D. Botstein. Cluster analysis and display of genome-wide expression patterns. *Proceedings of the National Academy of the Sciences USA*, 95:14863–14868, 1998.

C.T. Ekstarøm and P. Dalgaard. Linkage analysis of quantitative trait loci in the presence of heterogeneity. *Human Heredity*, 55:16–26, 2003.

F.V. Elmsie, M. Rees, M.P. Williamson, M. Kerr, M.J. Kjeldsen, K.A. Pang, et al. Genetic mapping of a major susceptibility locus for juvenile epilepsy on chromosome 15q. *Human Molecular Genetics*, 6:1329–1334, 1997.

R.C. Elston and J. Stewart. A general model for the analysis of pedigree data. *Human Heredity*, 21:523–542, 1971.

J.L. Fleiss. *The Design and Analysis of Clinical Experiments*. Wiley, New York, 1986.

R. Fletcher. *Practical Methods of Optimization*. John Wiley and Sons, Inc., New York, 1987.

Y. Freund and R.E. Schapire. A decision-theoretic generalization of on-line learning and an application to boosting. *Journal of Computer and System Sciences*, 55:119–139, 1997.

H.P. Friedman and J. Rubin. On some invariant criteria for grouping data. *Journal of the American Statistical Association*, 62:1159–1178, 1967.

J. Friedman, T. Hastie, and R. Tibshirani. Additive logistic regression: a statistical view of boosting (with discussion and a rejoinder by the authors). *The Annals of Statistics*, 28:337–407, 2000.

M. Frith, U. Hansen, and Z. Weng. Detection of *cis*-element clusters in higher eukaryotic DNA. *Bioinformatics*, 17:878–889, 2001.

S. Frühwirth-Schnatter. Data augmentation and dynamic linear models. *Journal of Time Series*

Analysis, 15:183–202, 1994.

D.W. Fulker, S.S. Cherny, and L.R. Cardon. Multipoint interval mapping of quantitative trait loci, using sib pairs. *American Journal of Human Genetics*, 56:1224–1233, 1995.

G. Gan, C. Ma, and J. Wu. *Data Clustering: Theory, algorithms and applications*. ASA-SIAM Series on Statistics and Applied Probability, SIAM, Philadelphia, ASA, Alexandria, VA, 2007.

A.P. Gasch, P.T. Spellman, C.M. Kao, O. Carmel-Harel, et al. Genomic expression programs in the response of yeast cells to environmental changes. *Molecular Biology of the Cell*, 11:4241–4257, 2000.

W.J. Gauderman and D.C. Thomas. Censored survival models for genetic epidemiology: A Gibbs sampling approach. *Genetic Epidemiology*, 11:171–188, 1994.

M.S. Gelfand, A.A. Mironov, and P.A. Pevzner. Gene recognition via spliced sequence alignment. *Proceedings of the National Academy of Sciences USA*, 93:9061–9066, 1996.

A. Gelman, J. Carlin, H. Stern, and D. Rubin. *Bayesian Data Analysis*. Chapman and Hall, New York, 2004.

A. Gelman and D.B. Rubin. Inference from iterative simulation using multiple sequences (with discussion). *Statistical Science*, 7:457–511, 1992.

W. Gish and D.J. States. Identification of protein coding regions by database similarity search. *Nature Genetics*, 3:266–272, 1993.

B. Gnedenko. Sur la distribution limite du terme maximum d'une serie aleatoire. *Ann. Math.*, 44:423–453, 1943.

D.E. Goldgar. Multipoint analysis of human quantitative genetic variation. *American Journal of Human Genetics*, 47:957–967, 1990.

T.R. Golub, D.K. Slonim, P. Tamayo, et al. Molecular classification of cancer: Class discovery and class prediction by gene expression monitoring. *Science*, 286:531–537, 1999.

V. Gosh Tusher, R. Tibsharani, and G. Chu. Significance analysis of microarrays applied to the ionizing radiation response. *Proceedings of the National Academy of the Sciences USA*, 98:5116–5121, 2001.

J.H. Graber, C.R. Cantor, S.C. Mohr, and T.F. Smith. *In silico* detection of control signals: mRNA 3'-end-processing sequences in diverse species. *Proceedings of the National Academy of Sciences USA*, 96:14055–14060, 1999.

D.M. Green and J.M. Swets. *Signal detection theory and psychophysics*. John Wiley and Sons Inc., New York, 1966.

W.N. Grundy, T.L. Bailey, C.P. Elkan, and M.E. Baker. Meta-MEME: Motif-based hidden Markov models of protein families. *Computer Applications in the Biosciences*, 13:397–406, 1997.

R. Guigó, P. Agarwal, J.F. Abril, et al. An assessment of gene prediction accuracy in large DNA sequences. *Genome Research*, 10:1631–1642, 2000.

S-W Guo. Computation of identity-by-descent proportions shared by two siblings. *American Journal of Human Genetics*, 54:1104–11109, 1994.

S-W Guo. Linkage disequilibrium measures for fine-scale mapping: A comparison. *Human Heredity*, 47:301–314, 1997.

J.B.S. Haldane. The combination of linkage values and the calculation of distances between the loci of linked factors. *Journal of Genetics*, 8:299–309, 1919.

S. Hannenhalli and L.S. Wang. Enhanced position weight matrices using mixture models. *Bioinformatics*, 23:i204–i212, 2005.

J.A. Hartigan. Direct clustering of a data matrix. *Journal of the American Statistical Association*, 67:123–129, 1972.

J.A. Hartigan. *Clustering Algorithms*. Wiley, New York, 1975.

J.K. Haseman and R.C. Elston. The investigation of linkage between a quantitative trait and a marker locus. *Behavioral Genetics*, 2:3–19, 1972.

T. Hastie, R. Tibshirani, and J. Friedman. *The Elements of Statistical Learning: Data Mining, Inference, and Prediction*. Springer, New York, 2001.

S.C. Heath. Markov chain Monte Carlo segregation and linkage analysis for oligogenic models. *American Journal Human Genetics*, 61:748–760, 1997.

J.G. Henikoff and S. Nehikoff. Using substitution probabilities to improve position specific scoring matrices. *Computer Applications in the Biosciences*, 12:135–143, 1996.

S.G. Hilsenbeck, W.E. Friedrichs, R. Schiff, et al. Statistical analysis of array expression data as applied to the problem of tamoxifen resistance. *Journal of the National Cancer Institute*, 91:453–459, 1999.

S. Hoffjan, D. Nicolae, and C. Ober. Association studies for asthma and atopic diseases: A comprehensive review of the literature. *Respiratory Research*, 3:4–14, 200.

R. Hoffman, T. Seidl, and M. Dugas. Profound effect of normalization on detection of differentially expressed genes in oligonucletide microarray data analysis. *Genome Biology*, 3:33.1–33.11, 2002.

N.S. Holter, M. Mitra, A. Maritan, M. Cieplak, J.R. Banavar, and N.V. Fedoroff. Fundamental patterns underlying gene expression profiles: Simplicity from complexity. *Proceedings of the National Academy of the Sciences USA*, 97:8409–8414, 2000.

S. Horvath, X. Xu, and N. Laird. The family based association test method: Strategies for studying general genotype-phenotype associations. *European Journal of Human Genetics*, 9:301–306, 2001.

J.G. Ibrahim, M-H Chen, and R.J. Gray. Bayesian models for gene expression with DNA microarray data. *Journal of the American Statistical Association*, 97:88–99, 2002.

R.A. Irizarry, B. Hobbs, F. Collin, Y.D. Beazer-Barclay, K.J. Antonellis, U. Scherf, and T.P. Speed. Exploration, normalization, and summaries of high density oligonucleotide array probe level data. *Biostatistics*, 4:249–264, 2003.

S.J. Iturria and J. Blangero. An EM algorithm for obtaining maximum likelihood estimates in the multi-phenotype variance components linkage model. *Annals of Human Genetics*, 64:349–362, 2000.

S. Karlin and S.F. Altschul. Methods for assessing the statistical significance of molecular sequence features by using general scoring schemes. *Proceedings of the National Academy of Sciences USA*, 87:2264–2268, 1990.

L. Kaufman and P. Rousseeuw. *Finding Groups in Data-An Introduction to Cluster Analysis*. Wiley, New York, 1990.

M.J. Kearns and U.V. Vazirani. *An Introduction to Computational Learning Theory*. MIT Press, Boston, MA, 1994.

K. Keffaber. Reproductive failure of unknown etiology. *American Association of Swine Practitioners Newsletter*, pages 1–9, 1989.

G. Kitigawa. Non-Gaussian state-space modeling of nonstationary time series. *Journal of the American Statistical Association*, 82:1032–1041, 1987.

Y. Kluger, R. Basri, J.T. Chang, and M. Gerstein. Spectral biclustering of microarray data: Coclustering genes and conditions. *Genome Research*, 13:703–716, 2003.

M. Knapp. The transmission/disequilibrium test and parental-genotype reconstruction: The reconstruction-combined transmission/disequilibrium test. *American Journal of Human Genetics*, 64:861–870, 1999.

K. Koinuma, Y. Yamashita, W. Liu, H. Hatanaka, et al. Epigenetic silencing of axin2 in colorectal carcinoma with microsatellite instability. *Oncogene*, 25:139–146, 2006.

A. Krogh, M. Brown, I.S. Mian, K. Sjölander, and D. Haussler. Hidden Markov models in computational biology: Application to protein modeling. *Journal of Molecular Biology*, 235:1501–1531, 1994.

W.J. Krzanowski and Y.T. Lai. A criterion for determining the number of groups in a data set using sum of squares clustering. *Biometrika*, 44:23–34, 1985.

D. Kulp, D. Haussler, M.G. Reese, and F.H. Eeckman. A generalized hidden Markov model for the recognition of human genes in DNA. *Proceedings of the International Conference on Intelligent Systems in Molecular Biology*, 4:134–142, 1996.

E. Lander and P. Green. Construction of multilocus genetic maps in humans. *Proceedings of the National Academy of Sciences USA*, 84:2363–2367, 1987.

E.S. Lander and M.S. Waterman. Genomic mapping by fingerprinting random clones: A mathematical analysis. *Genomics*, 2:231–23, 1988.

K. Lange. *Mathematical and Statistical Methods for Genetic Analysis*. Springer, New York, NY, 1997.

K. Lange and M. Boehnke. Bayesian methods and optimal experimental design for gene mapping by radiation hybrids. *Annals of Human Genetics*, 56:119–144, 1992.

G. Lathrop, J. Lalouel, C. Julier, and J. Ott. Multilocus linkage analysis in humans: Detection of linkage and estimation of recombination. *American Journal of Human Genetics*, 37:482–498, 1985.

C.E. Lawrence and A.A. Reilly. An expectation maximization algorithm for the identification and characterization of common sites in unaligned biopolymer sequences. *PROTEINS: Structure, Function and Genetics*, 7:41–51, 1990.

J.K. Lee, C. Park, K. Kimm, and M.S. Rutherford. Genome-wide multilocus analysis for immune-mediated complex diseases. *Biochemical Biophysical Research Communications*, 295:771–773, 2002.

M. Levitt and M. Gerstein. A unified statistical framework for sequence comparison and structure comparison. *Proceedings of the National Academy of Sciences USA*, 95:5913–5920, 1998.

S. Levy and S. Hannenhalli. Identification of transcription factor binding sites in the human genome sequence. *Mammalian Genome*, 13:510–514, 2002.

C. Li and W.H. Wong. Model-based analysis of oligonucleotide arrays: Expression index computation and outlier detection. *Proceedings of the National Academy of the Sciences USA*, 98:31–36, 2001.

C. Li and W.H. Wong. Model-based analysis of oligonucleotide arrays: Model validation, design issues and standard error application. *Genome Biology*, 2:32.1–32.11, 2001.

W. Li, C.S.J. Fann, and J. Ott. Low-order polynomial trends of female-to-male map distance ratios along human chromosomes. *Human Heredity*, 48:266–270, 1998.

U. Liberman and S. Karlin. Theoretical models of genetic map functions. *Theoretical Population Biology*, 25:331–346, 1984.

R.J. Lipshutz, S.P. Fodor, T.R. Gingeras, and D.J. Lockhart. High density synthetic oligonucleotide arrays. *Nature Genetics (Supplement 1)*, 21:20–24, 1998.

B.H. Liu. *Statistical Genomics*. CRC Press, Boca Raton, 1998.

J.S. Liu. The collapsed Gibbs sampler in Bayesian computations with applications to a gene regulation problem. *Journal of the American Statistical Association*, 89:958–966, 1994.

J.S. Liu, A. Neuwald, and C. Lawrence. Bayesian models for multiple local sequence alignment and Gibbs sampling strategies. *Journal of the American Statistical Association*, 90:1156–1170, 1995.

J.S. Liu, A. Neuwald, and C. Lawrence. Markovian structures in biological sequence alignments. *Journal of the American Statistical Association*, 94:1–15, 1999.

H. Lodish, A. Berk, P. Matsudaira, D. Baltimore, and J. Darnell. *Molecular Cell Biology*.

Freeman, 2001.

K.L. Lunetta, M. Boehmke, K. Lange, and D.R. Cox. Selected locus and multiple panel models for radiation hybrid mapping. *American Journal of Human Genetics*, 59:717–725, 1996.

J.B. MacQueen. Some methods for classification and analysis of multivariate observations. In *Proceedings of the 5th Berkeley Symposium on Mathematical Statistics and Probability*, volume 1, pages 281–297, Berkeley, CA, 1967. University of California Press.

O.L. Mangarasian. *Nonlinear Programming*. McGraw Hill, New York, 1969.

V. Matys, O.V. Kel-Margoulis, E. Fricke, I. Liebich, S. Land, A. Barre-Dirrie, et al. TRANS-FAC and its module TRANSCompel: Transcriptional gene regulation in eukaryotes. *Nucleic Acids Research*, 34:D108–D110, 2006.

J.D. McAuliffe, L. Pachter, and M.I. Jordan. Multiple-sequence functional annotation and the generalized hidden Markov phylogeny. *Bioinformatics*, 20:1850–1860, 2004.

G.P. McCormick. *Non Linear Programming: Theory, Algorithms and Applications*. John Wiley and Sons, Inc., New York, 1983.

I. Meyer and R. Durbin. Comparative *ab initio* prediction of gene structures using pair HMMs. *Bioinformatics*, 18:1309–1318, 2002.

H. Mi, B. Lazareva-Ulitsky, R. Loo, et al. The PANTHER database of protein families, subfamilies, functions and pathways. *Nucleic Acids Research*, 33:D284–D288, 2005.

G.W. Milligan and M.C. Cooper. An examination of procedures for determining the number of clusters in a data set. *Psychometrika*, 50:159–179, 1985.

K.C. Moraes, C.J. Wilusz, and J. Wilusz. CUG-BP binds to RNA substrates and recruits PARN deadenylase. *RNA*, 12:1084–1091, 2006.

J.J. Morè and S.J. Wright. *Optimization Guide*. SIAM, 1993.

N.E. Morton. The detection and estimation of linkage between genes for elliptocytosis and Rh blood type. *American Journal of Human Genetics*, 8:80–96, 1956.

M.A. Newton, C.M. Kendziorski, C.S. Richmond, F.R. Blattner, and K.W. Tsui. On differential variability of expression ratios: Improving statistical inference about gene expression changes from microarray data. *Journal of Computational Biology*, 8:37–52, 2001.

J. Norris. *Markov Chains*. Cambridge University Press, UK, 1997.

J.R. O'Connell and D.E. Weeks. Pedcheck: A program for identifying genotype incompatibilities in linkage analysis. *American Journal Human Genetics*, 63:259–266, 1998.

M. Oliver, A. Aggarwal, J. Allen, et al. A high-resolution radiation hybrid map of the human genome draft sequence. *Science*, 291:1298–1302, 2001.

The Collaborative Study on the Genetics of Asthma (CSGA). A genome-wide search for asthma susceptibility loci in ethnically diverse populations. *Nature Genetics*, 15:389–392, 1997.

R. Osada, E. Zaslavsky, and M. Singh. Comparative analysis of methods for representing and searching for transcription factor binding sites. *Bioinformatics*, 20:3516–3525, 2004.

J. Ott. Linkage analysis and family classification under heterogeneity. *Annals of Human Genetics*, 47:311–320, 1983.

J. Ott. Statistical properties of the haplotype relative risk. *Genetic Epidemiology*, 6:127–130, 1989.

J. Ott. *Analysis of Human Genetic Linkage*. Johns Hopkins Press, Baltimore, 1999.

G. Pavesi, M. Giancarlo, and G. Pesole. An algorithm for finding signals of unknown length in DNA sequences. *Bioinformatics*, 17:S207–S214, 2001.

J. Pederson and J. Hein. Gene finding with a hidden Markov model of genome structure and evolution. *Bioinformatics*, 19:219–227, 2003.

L. Penrose. The detection of autosomal linkage in data which consists of pairs of brothers and sisters of unspecified parentage. *Annals of Eugenics*, 6:133–138, 1935.

D. Pohlers, A. Beyer, D. Koczan, T. Wilhelm, et al. Constitutive upregulation of the transforming growth factor-beta pathway in rheumatoid arthritis synovial fibroblasts. *Arthritis Research and Therapy*, 9:R59, 2007.

D. Porter and D.S.G. Stirling. *Integral Equations*. Cambridge University Press, Cambridge, 1990.

W. Press, S. Teukolsky, W. Vetterling, and B. Flannery. *Numerical Recipes in C*. Cambridge University Press, UK, 1992.

D. Prestidge. Predicting Pol II promoter sequences using transcription factor binding sites. *Journal of Molecular Biology*, 249:923–932, 1995.

A.L Price, N.J. Patterson, B.M. Plenge, M.E. Weinblatt, N.A. Schadick, and D. Reich. Principle components analysis corrects for stratification in genome-wide association studies. *Nature Genetics*, 38:904–909, 2006.

J.K. Pritchard, M. Stephens, and P. Donnelly. Inference of population structure using multilocus data. *Genetics*, 155:945–959, 2000.

K. Quandt, K. Frech, H. Karas, E. Wingender, and T. Werner. MatInd and MatInspector: New fast and versatile tools for detection of consensus matches in nucleotide sequence data. *Nucleic Acids Research*, 23:4878–4884, 1995.

D. Rabinowitz. A transmission disequilibrium test for quantitative trait loci. *Human Heredity*, 47:342–350, 1997.

A. Raghavan, R.L. Ogilvie, C. Reilly, M.L. Abelson, S. Raghavan, J. Vasdewani, et al. Genome-wide analysis of mRNA decay in resting and activated primary human T lymphocytes. *Nucleic Acids Research*, 30:5529–5538, 2002.

J.T. Reese and W.R. Pearson. Empirical determination of effective gap penalties for sequence comparison. *Bioinformatics*, 11:1500–1507, 2002.

C. Reilly, A. Gelman, and J. Katz. Post-stratification without population level information on the post-stratifying variable, with application to political polling. *Journal of the American Statistical Association*, 96:1–11, 2001.

C. Reilly, M.B. Miller, Y. Liu, W.S. Oetting, R. King, and M. Blumenthal. Linkage analysis of a cluster-based quantitative phenotype constructed from pulmonary function test data in 27 multigenerational families with multiple asthmatic members. *Human Heredity*, 64:136–145, 2007.

C. Reilly, C. Wang, and M. Rutherford. A method for normalizing microarrays using the genes that are not differentially expressed. *Journal of the American Statistical Association*, 98:868–878, 2003.

J.A. Rice. *Mathematical Statistics and Data Analysis*. Duxbury Press, Belmont, CA, 1995.

M.J. Robinson and M.H. Cobb. Mitogen activated kinase pathways. *Current Opinion in Cell Biology*, 9:180–186, 1977.

K. Rossow, J. Shivers, P. Yeske, D. Polson, R. Rowland, S. Lawson, et al. Porcine reproductive and respiratory syndrome virus infection in neonatal pigs characterized by marked neurovirulence. *Veterinary Research*, 144:444–448, 1999.

M. Sagot. Spelling approximate repeated or common motifs using a suffix tree. *Lecture Notes in Computer Science*, 1380:111–127, 1998.

A. Sandelin, W. Alkema, P. Engström, W.W. Wasserman, and B Lenhard. JASPAR: an open-access database for eukaryotic transcription factor binding profiles. *Nucleic Acids Research*, 32:D91–D94, 2004.

E.E. Schadt, C. Li, C. Su, and W.H. Wong. Analyzing high-density oligonucleotide gene expression array data. *Journal of Cellular Biochemistry*, 80:192–202, 2000.

D. Scott. *Multivariate Density Estimation: Theory, Practice, and Visualization*. Wiley, New York, 1992.

D. Shmulewitz, S.C. Heath, M.L. Blundell, Z. Han, R. Sharma, J. Salit, et al. Linkage analysis of quantitative traits for obesity, diabetes, hypertension, and dyslipidemia on the island of Kosrae, Federated States of Micronesia. *Proceedings of the National Academy of the Sciences USA*, 103:3502–3509, 2006.

B.W. Silverman. *Density Estimation for Statistics and Data Analysis*. Chapman & Hall, Boca Raton, FL, 1986.

S.D. Silvey. *Statistical Inference*. Chapman & Hall, London, 1975.

C. Smith. Some comments on the statistical methods used in linkage investigations. *American Journal of Human Genetics*, 11:289–304, 1959.

V.V. Solovyev, A.A. Salamov, and C.B. Lawrenece. Predicting internal exons by oligonucleotide composition and discriminant analysis of spliceable open reading frames. *Nucleic Acids Research*, 22:248–250, 1994.

R.S. Spielman, R.E. McGinnis, and W.J. Ewens. Transmission test for linkage disequilibrium: The insulin gene region and insulin-dependent diabetes mellitus. *American Journal of Human Genetics*, 52:506–516, 1993.

J.D. Storey. A direct approach to false discovery rates. *Journal of the Royal Statistical Society, B*, 64:479–498, 2002.

J.D. Storey. The positive false discovery rate: a Bayesian interpretation and the q-value. *Annals Statistics*, 31:2013–2035, 2003.

J.D. Storey and R. Tibshirani. Statistical significance for genomewide studies. *Proceedings of the National Academy of the Sciences USA*, 100:9440–9445, 2003.

T. Strachan and A. Read. *Human Molecular Genetics*. Garland Science, London, 2004.

C. Terpstra, G. Wensvoort, and J. Pol. Experimental reproduction of porcine epidemic abortion and respiratory syndrome (mystery swine disease) by infection with Lelystad virus: Koch's postulates fulfilled. *Veterinary Quarterly*, 13:131–136, 1991.

J.D. Terwilliger and J. Ott. A haplotype-based haplotype relative risk approach to detecting allelic associations. *Human Heredity*, 42:337–346, 1992.

G. Thijs, M. Lescot, K. Marchal, S. Rombauts, B. De Moor, P. Rouzé, and Y. Moreau. A higher-order background model improves the detection of promoter regulatory elements by Gibbs sampling. *Bioinformatics*, 17:1113–1122, 2001.

M. Thompson, R. McInnes, and H. Willard. *Genetics in Medicine*. W.B. Saunders Co., Philadelphia, 1991.

W. Thompson, E.C. Rouchka, and C.E. Lawrence. Gibbs recursive sampler: Finding transcription factor binding sites. *Nucleic Acids Research*, 31:3580–3585, 2003.

R. Tibsharani, G. Walther, and T. Hastie. Estimating the number of clusters in a dataset via the gap statistic. *Journal of the Royal Statistical Society, B*, 63:411–423, 2001.

R. Tibshirani. Regression shrinkage and selection via the lasso. *Journal of the Royal Statistical Society, B.*, 58:267–288, 1996.

M-L. Ting Lee, F.C. Kuo, G.A. Whitmore, and J. Sklar. Importance of replication in microarray gene expression studies: Statistical methods and evidence from repetitive cDNA hybridizations. *Proceedings of the National Academy of the Sciences USA*, 97:9834–9839, 2000.

M. Tompa, N. Li, T.L. Bailey, G.M. Church, B. De Moor, E. Eskin, et al. Assessing computational tools for the discovery of transcription factor binding sites. *Nature Biotechnology*, 23:137–144, 2005.

E.H. Trager, R. Khanna, A. Marrs, L. Siden, K.E.H. Branham, A. Swaroop, et al. Madeline 2.0 PDE: A new program for local and web-based pedigree drawing. *Bioinformatics*, 23:1854 – 1856, 2007.

G. Tseng, M. Oh, L. Rohlin, J. Liao, and W. Wong. Issues in cDNA microarray analysis: quality filtering, channel normalization, models of variations and assessment of gene effects. *Nucleic Acids Research*, 29:2549–2557, 2001.

L.G. Valiant. A theory of the learnable. *Communications of the ACM*, 27:1134–1142, 1984.

V. Vapnik. *Estimation of Dependences Based on Empirical Data [English translation]*. Springer Verlag, New York, 1982.

V. Vapnik. *The Nature of Statistical Learning Theory*. Springer-Verlag, New York, 1995.

I.A. Vlasova, N.M. Tahoe, D. Fan, O. Larsson, B. Rattenbacher, J.R. Sternjohn, J. Vasdewani, G. Karypis, C.S. Reilly, P.B. Bitterman, and P.R. Bohjanen. Conserved GU-rich elements mediate mRNA decay by binding to CUG-binding protein 1. *Molecular Cell*, 29:263–270, 2008.

F. Vogel and A. Motulsky. *Human Genetics*. Springer, Berlin, 1997.

M.G. Wade, A. Kawata, A. Williams, and C. Yauk. Methoxyacetic acid-induced spermatocyte death is associated with histone hyperacetylation in rats. *Biology of Reproduction*, 78:822–831, 2008.

H. Wakaguri, R. Yamashita, Y. Suzuki, and K. Nakai. DBTSS: Database of transcriptional start sites, progress report 2008. *Nucleic Acids Research*, 36:D97–101, 2008.

C. Wang, R. Hawken, E. Larson, X. Zhang, L. Alexander, and Rutherford M. Generation and mapping of expressed sequence tags from virus-infected swine macrophages. *Animal Biotechnology*, 12:51–67, 2001.

N. Wang, K.K. Lin, Z. Lu, K.S. Lam, et al. The lim-only factor lmo4 regulates expression of the bmp7 gene through an hdac2-dependent mechanism, and controls cell proliferation and apoptosis of mammary epithelial cells. *Oncogene*, 26:6431–6441, 2007.

J.H. Ward. Hierarchical groupings to optimize an objective function. *Journal of the American Statistical Association*, 58:236–244, 1963.

W. Wasserman, M. Palumbo, W. Thompson, J. Fickett, and C. Lawrence. Human-mouse genome comparisons to locate regulatory sites. *Nature Genetics*, 26:225–228, 2000.

M.S. Waterman. *Introduction to Computational Biology*. Chapman & Hall/CRC, Boca Raton, 1995.

B-J. Webb, J.S. Liu, and C.E. Lawrence. BALSA: Bayesian algorithms for local sequence alignment. *Nucleic Acids Research*, 30:1268–1277, 2002.

W. Weinberg. Mathematische gundlagen der probandenmethode. *Z Indukt Abstamm u Vererb Lehre*, 48:179–228, 1927.

X. Wen, S. Fuhrman, G.S. Michaels, D.B. Carr, S. Smith, J.L. Barker, and R. Somogyi. Large scale temporal gene expression mapping of central nervous system development. *Proceedings of the National Academy of the Sciences USA*, 95:334–339, 1998.

G. Wensvoort, C. Terpstra, J. Pol, E. ter Laak, M. Bloemraad, E. de Kluyver, C. Kragten, and et al. Mystery swine disease in the Netherlands: The isolation of Lelystad virus. *Veterinary Quarterly*, 13:121–130, 1991.

A. Whittemore and I. Tu. Simple, robust linkage tests for affected sibs. *American Journal of Human Genetics*, 62:1228–1242, 1998.

W.J. Wilbur and D.J. Lipman. Rapid similarity searches of nucleic acid and protein data banks. *Proceedings of the National Academy of Sciences USA*, 80:726–730, 1983.

J. Wittes and H.P. Friedman. Searching for evidence of altered gene expression: A comment on statistical analysis of microarray data. *Journal of the National Cancer Institute*, 91:400–401, 1999.

Z. Wu and R.A. Irizarry. Stochastic models inspired by hybridization theory for short oligonu-cleotide arrays. *Journal of Computational Biology*, 12:882–893, 2005.

Y. Yang, S. Dudoit, P. Luu, D. Lin, V. Peng, J. Ngai, and T.P. Speed. Normalization for cDNA microarray data: a robust composite method addressing single and multiple slide systematic variation. *Nucleic Acids Research*, 30:1–10, 2002.

Y. Yang, S. Dudoit, P. Luu, and T.P. Speed. Normalization for cDNA microarray data. In M. Bittner, Y. Chen, A. Dorsel, and E. Dougherty, editors, *Microarrays: Optical Technologies and Informatics*, volume 4266. SPIE, 2001.

K. Y. Yeung and W. L. Ruzzo. Principal component analysis for clustering gene expression data. *Bioinformatics*, 17:763–774, 2001.

K.Y. Yeung, C. Fraley, A. Murua, A.E. Raftery, and W.L. Ruzzo. Model-based clustering and data transformations for gene expression data. *Bioinformatics*, 17:977–987, 2001.

R.A. Young. Biomedical discovery with DNA arrays. *Cell*, 102:9–15, 2000.

L. Zhang, M. Miles, and K. Aldape. A model of molecular interactions on short oligonucleotide microarrays. *Nature Biotechnology*, 21:818–821, 2003.

M. Zhang. Computational prediction of eukaryotic protein-coding genes. *Nature Reviews*, 3:698–710, 2002.

M.Q. Zhang. Identification of protein coding regions in the human genome based on quadratic discriminant analysis. *Proceedings of the National Academy of Science USA*, 94:565–568, 1997.

M.Q. Zhang. Statistical features of human exons and their flanking regions. *Human Molecular Genetics*, 7:919–932, 1998.

X. Zhang, J. Shin, T. Molitor, L. Schook, and M. Rutherford. Molecular responses of macrophages to porcine reproductive and respiratory syndrome virus infection. *Virology*, 262:152–162, 1999.

L.P. Zhao, R. Prentice, and L. Breeden. Statistical modeling of large microarray data sets to identify stimulus-response profiles. *Proceedings of the National Academy of the Sciences USA*, 98:5631–5636, 2001.

Index

For Product Safety Concerns and Information please contact our EU
representative GPSR@taylorandfrancis.com Taylor & Francis Verlag GmbH,
Kaufingerstraße 24, 80331 München, Germany

Printed and bound by CPI Group (UK) Ltd, Croydon, CR0 4YY

01/05/2025

01858569-0001